MW00980248

To Paula

CONTACTO SILVESTRE *ediciones*
de Santiago G. de la Vega
Av. del Libertador 1396 - Piso 6º B
(1638) - Vicente López - Buenos Aires - Argentina
Ph / Fax: (54 11) 4795-1727
e-mail: sdelavega@contactosilvestre.com.ar
www.contactosilvestre.com.ar

IGUAZÚ, The LAWS of the JUNGLE
All rights reserved. No part of this publication may be reproduced, stored in a retrieval system or transmitted, in any form or by any means, without the prior written consent of the author and publisher.

I.S.B.N. 987-97402-8-9
Queda hecho el depósito que marca la Ley 11.723

Title in Spanish:
IGUAZÚ, Las LEYES de la SELVA
First published February 1999
First english edition - August 1999
Second english edition - June 2003

Illustrations of flora and fauna:
Gustavo R. Carrizo

Photography:
Jorge Schulte

Translation:
Deborah Primrose

Graphic Design:
Verónica Martorell

Printed in:
Gráfica LAF S.R.L.
Loyola 1654 (1414) Buenos Aires Argentina
Ph: (54 11) 4855-8171 - (54 11) 4854-3893
trazos@arnet.com.ar

Published June, 2003. Buenos Aires, Argentina.

IGUAZÚ
The LAWS of the JUNGLE

Santiago G. de la Vega

Illustrations: Gustavo R. Carrizo

Contributors:
Andrés Bosso, Aves Argentinas - Asociación Ornitológica del Plata
Juan Carlos Chebez, Administración de Parques Nacionales
Rosendo Fraga, Aves Argentinas - Asociación Ornitológica del Plata
Justo Herrera, Administración de Parques Nacionales

AVES ARGENTINAS®
Asociación Ornitológica del Plata

APN

FUNDACION
VIDA SILVESTRE
ARGENTINA

CONTACTO
SILVESTRE
e d i c i o n e s

www.contactosilvestre.com.ar

Acknowledgements

In chronological order, firstly I would like to thank the people who have influenced me in these last years, in a greater or lesser degree, to acquire the "ingredients" of knowledge and motivation towards Nature needed to write this book.

To Gustavo Ferreyra, of the Instituto Antártico Argentino, to Hernán Pueyrredon and Fermín Reinoso, of the Estancia Bouvier, in Formosa. They gave me the opportunity to live the most unforgettable experiences of "contacto silvestre" (contact with the wilderness), first in the Antarctic and in the Chaco environment later.
A special acknowledgement to Marlú and Federico B. Kirbus who incited me, since I met them in 1991, to write for several media, and invited me to travel to fantastic corners of the country. I would also like to thank Federico for correcting this book from the point of view of an experienced traveller and writer.
To Ricardo Clark, Miguel Tauszig, Andrés Hollman and Diego Naveiro, as Travel Agents, for giving me the opportunity to guide Nature trips in several areas of Argentina and Chile. To Christopher Cutler and Mark Smith, important drive sources, after sharing with them Nature trips around Argentina and Chile.
For complementing this written work with his "animated" illustrations of the flora and fauna, I would specially like to thank Gustavo Rodolfo Carrizo, a naturalist of the "Bernardino Rivadavia" Natural History Museum.

Once these pages were taking shape, the help of the following contributors was fundamental to enrich and improve its content:
Andrés Bosso, Juan Carlos Chebez and Rosendo Fraga, for their invaluable help in giving specific data and concepts, suggestions and their technical correction of the book.
Justo Herrera, who gave the location of the trees on the maps of the Waterfalls Circuits contributed with data and concepts about the jungle "in situ" and the technical correction of botanical subjects.
I must also express my gratitude to the help given by Marcelo Canevari, of the Administración de Parques Nacionales; and to Hugo Chaves. Sofía Heinonen Fortabat and Ariel Soria, members of the Delegación Técnica Regional del Noreste, and Silvina Fabri, of the Centro de Investigaciones Ecológicas Subtropicales (CIES).

On the other hand, the bibliography has been essential. In such a respect, I would like to thank Eugenio Coconier, of the Asociación Ornitológica del Plata's Library; Laura Malmieca, of the CIES, for allowing me to consult bibliography in this Centre. I would also like to thank the staff of the library of the Instituto Darwinion and of the Administración de Parques Nacionales; and Colin Sharp, of the bookstore L.O.L.A. The books referred to represent, in most cases, samples of years of work of biologists and naturalists of Argentina and other countries dedicated to research and/or conservation and divulgence of Nature.

I would like to thank Cristian Henshcke, Diego Gallegos Luque and Daniel Somay for their general revision. Among the people with the perception of "visitor of Iguazú", I would like to thank my brother Diego de la Vega, my father Carlos Alberto de la Vega, Jose Roque Laurnagaray, Paula Levallois and Lucía Molteni for their comments and partial or total revision of the text. To the staff of Casa de la Provincia de Misiones.
Special thanks to Jorge Shulte, for providing his photographs, and Deborah Primrose, for all her work with the translation.

In reference to CONTACTO SILVESTRE ediciones, started with the aim of publishing topics related to Nature, I would like to thank Fernando María Borghesi and Flavio De Pietro from "El Bloque" design studio, for their help in the first stage. I would also like to thank Hector Aranda, from Clarín, Miguel Capuz and Silvia Varela, from the Capuz-Varela Editions, Cecilia Luchía Puig, editor of the magazine Mañana Profesional, and my brothers Juan María and Javier de la Vega.

Finally, a special acknowledgement to Verónica Martorell, who worked the most to make this publication true. The creative graphic design of this small book is only a visible sample of all her work and collaboration.

To all, thank you very much.

Santiago G. de la Vega
biologist
avid of Nature

... **Swifts** (family Apodidae) are probably the birds that spend most of their life flying, to such an extent that they hunt, eat, obtain material for their small nests and can even mate in the air. On the other hand, their legs are so short, "adapted to hang from surfaces", and their folded wings are so long that they can hardly stand on the ground (see page 82).

reat Dusky vift

... the iridescent metallic blue colour of the *Morpho* **butterflies** would mean to their predators a message that says "I accelerate too fast to be caught". The stunning design of colours of the butterflies *Heliconius* warns "Don't try to eat me, I am toxic" (see page 44).

Morpho

Heliconius

... **hummingbirds** (family Trochilidae) remain suspended in the air (hovering flight) to feed on nectar from flowers and are the only birds capable of flying backwards thanks to the combination of their small size, the shape and the high frequency beating of their wings. In the Amethystine Hummingbird, a species which is rare in Iguazú, the beating of wings can reach up to 80 beats per second.
You have just read some interesting data, now close your eyes and try to imagine: 80 beats per second! (see page 87).

Black-breasted Plovercrest

... **hawkmoths** (family Sphingidae) also feed on the nectar of flowers while flying and can beat their wings with a frequency of up to 200 times per second, with a movement of wings similar to that of the hummingbirds (see page 67). But, while butterflies move around during daytime and are attracted specially by red and orange flowers in search of nectar, hawkmoths

Sphingidae hawkmoth

fly in the dark captivated by the fragrances liberated by light coloured flowers that open during the night (see page 27).

... **Orchids** are probably the family of flowering plants with a greater diversity of species and specialisation in the world.

Orchid Miltonia flavescens

In many of the species one of the petals of the flower is of a greater size and has a very lively colour, as in *Miltonia flavescens*, common in Misiones. This petal represents a "marked landing site" for the attracted insects, which act as a "door to door mail" as they transport pollen that sticks to their body from one flower to another, thus ensuring the union with the ovule and so the formation of seeds (see page 26).

What do the insects get in exchange for their work?
Orchids offer different "prizes", although many species play tricks: there are cases (not found in Argentina) in which parts of the flower

resemble in shape and colour the females of certain wasps and bees to sexually attract the males (see page 28).

... the **Great Kiskadee** and its and relatives (tyrants) eat insects and to be more efficient in their hunting, focus their attention in "signals" of colours, shapes and movements of their probable preys. In this way they filter information that would otherwise distract them (see page 86).
To have a clearer idea of the concept: have you noticed that in the sentence where I mention the Great Kiskadee, there is an extra "and", or had you filtered it? If you did notice it, it is probably because you have a "prey" in your hands and you are carefully scrutinising it.

Great Kiskadee

On the other hand

Do you know that you have the opportunity of being amazed with these and other topics of the surprising natural world of the Iguazú National Park?

To help you in this task this book wants to show you in a simple and attractive way the interpretation of the "laws of the jungle" and the relationships between some plants and animals which can be observed in the Iguazú Waterfalls Circuits.

Many coloured illustrations complement the text and help in the identification of trees, hanging plants, butterflies, birds, mammals and several forms of life which you might find in your walks.

At the end of the book there are maps of the Waterfalls Circuits which indicate the location of the most important flora which is mentioned in the text. The "Seasonal Fruits" chart allows the reader to focus his attention on certain plants.

IGUAZÚ, the LAWS of the JUNGLE is focused on stimulating your curiosity toward Nature.

Till we are adolescents we are more curious, without the need of measuring utility or cost. It is said that people who keep this characteristic throughout the years are more creative and happier.

Now, whatever your age is, what is your **curiosity - curiosity - C-U-R-I-O-S-I-T-Y like?**

Stay
in close contact with Nature.

The only hotel located right by the falls is as spectacular as the surrounding landscape.

180 guestrooms and 4 suites. Swimming pool, tennis courts, play-room, gym and sauna. Recreational activities for all ages. International and regional specialties at our restaurants. Fully-equipped meeting and function rooms. Business Center with support secretarial services and meeting rooms. Starting point for walking tour of the falls...

Sheraton International Iguazú Resort, an unforgettable experience in a wondrous natural setting.

For further information and reservations:
Parque Nacional Iguazú,
3370 Iguazú, Prov. de Misiones, Argentina.
Tel.: (54) (3757) 491800.
Fax: (54) (3757) 491810.
Toll Free: 0800-888-9180.
Visit sheraton.com

Sheraton
Internacional Iguazú
RESORT

See for yourself

MEMBER OF STARWOOD PREFERRED GUEST

Index

Fauna

CONTACTO SILVESTRE
ediciones

Serie - Explorando Nuestra Naturaleza
to be published in English as
Argentina - Natural History Handbooks

www.contactosilvestre.com.ar

The rules of the game: How to use "The LAWS..."

The experience which you have as you enter the jungle without any knowledge about its inhabitants and how it "works", could be compared to that of practising a sport without knowing the "rules of the game". The best move might occur in front of your eyes without you noticing it.

If you think I'm exaggerating, you will realise what I mean as you read through these pages.

"The LAWS.." wants to show you the "rules of the game" with a simple and pleasant style. There are comments and comparisons with concepts easy to visualise which help to fix ideas.
As you enrich your personal experiences in Iguazú, you will become aware of the fantastic complexity of the environment.

This book is prepared to be used before, during and after the walks in the jungle in the following way:

Before:
• Go through the *Presentation* and *The LAWS of the JUNGLE*.

In the text you will find references which are connected with the topic to be seen with the indication of the page where to find the information.

During:
• Free your curiosity (unless you have it on a leash, in that case, let it loose!).
• The topics of the section *On the paths of Iguazú* will be your main guide.
• Consult the Maps of location of flora and the Seasonal Fruits Chart as you walk along the circuits.

Indications of the page numbers in the references of each map guide you to the corresponding topics.

After:
• With your observations in situ it is advisable to go through the different topics again. With more time to think, you will be able to report your experiences in a better way.
• At the end think about the section *The Misiones Jungle*.

Smooth-billed Ani and Cecropia Tree

Attitudes in the "playing field"

• The ABC for your walks: an "alert mind", "open eyes" and "ears on end". You should also have a "ready nose", and for the search of fauna: "Shhh..." (silence).

• You should be expecting to see not only large animals. The surprises could be in the smallest details.

• To observe flowers, fruits, seeds or leaves, don't cut them from the plants. You can pay attention to the samples which have fallen on the ground as if they were "pieces of crystal". They might still be food for many animals, and even give origin to new plants. Remember that hundreds of thousands of people visit this National Park each year.

• Butterflies are the insects which will probably call your attention most. It is possible to observe them as they fly or when they pose on flowers, leaves, puddles and even your own clothes or skin. Do not catch them. Likewise for any other animal. You should not feed wild animals.

• The paths are designed so you can enjoy the Waterfalls and the jungle during your walks.

• You must take care not to place your hands anywhere. You must prevent contact with poisonous animals, such as snakes or some species of spiders and insects.

Recommendations for the Walks

• Dawn and dusk are the best moments to explore the paths, specially for the observation of birds.

• Any time of the day is ideal to see plants or insects in general. You will also find several birds, lizards and who knows what other surprises...
Binoculars are indispensable for the birdwatching, although in Iguazú there are some species which can be seen without them.

• A magnifying glass helps to appreciate details and to discover a new world, specially with insects. To observe the object of interest with your binoculars in an inverted form is a way of improvising a magnifying glass.

• A notebook will become very handy to describe and record topics that interest you.

• You should not forget to take sun-screen lotion and insect repellent.

The "playing field":

Iguazú National Park:
Established in 1934.

Total Area:	67,000 hectares
National Reserve:	12,000 hectares
National Park:	55,000 hectares

It protects the Parananense Subtropical Forest and the Iguazú Waterfalls.
The Iguazú Waterfalls are a World Heritage Site (UNESCO, 1984).

What is the difference between a Subtropical and a Tropical Forest?
Let's make some comparisons between Iguazú and the Tropical Forest of the Barro
Colorado Island in Panama, Central America (one of the most studied jungles in
the world).

	Iguazú	Barro Colorado
Climate:	Subtropical, rainy, without a dry season.	Tropical, rainy, with a dry season.
Annual Average Rainfall:	2,000 mm	2,600 mm (90% during the rainy season).
Atmospheric humidity, annual average:		
Minimum:	74 %	68 %
Maximum:	85 %	93 %
	These averages are always greater near the Waterfalls.	
Average Temperatures:		
Annual	19 °C	27 °C
Monthly in winter:	15 °C	26 °C
Monthly in summer:	25 °C	28 °C
Solar Radiation, seasonal change:	variable	hardly variable

Comparing the Tropical Forest with Iguazú, the latter has:
A greater number of species of trees that lose their leaves seasonally (nearly a 50% of the species).
This is consequence of solar radiation decrease, and the lowering of temperatures during winter.
More adaptations to great changes in temperature between summer and winter, which gives it unique characteristics.
Less diversity of plants and animals.
A smaller size of trees.

The Iguazú jungle is a bordering area of distribution of several plant and animal groups which are more common in the Tropical Forest. This means a greater genetic variability. It has been said that genetic variability is to Nature as creativity is to man's culture.

The area closest to the Waterfalls, with the presence of water vapour, determines a special microclimate which favours the existence of species which are exclusive of that area. This is the case of many orchids (see page 60). The Cupay tree dominates the top stratum of the islands of the Upper Iguazú River, representing one of the few areas of distribution of this tree in Argentina. During the floods the water rises one or two meters above the normal level of these islands. There are some trees adapted to floods.

The Strata in
the Iguazú Jungle

Jungles have different strata. Vines and lianas interconnect the different strata and many plants grow on others in search of light.

The soil is rich in aluminium and iron and the humus layer is not very deep. Leaves, fallen branches, excrement, and all organic matter on the soil is decomposed quickly to inorganic compounds by bacteria and fungi.

Inorganic compounds re-enter the cycle through the roots, many of which have a relationship of mutual benefit with certain fungi (mycorrhizae). The roots benefit with more efficient absorption of water and minerals and the fungi obtain organic matter.

Trees and hanging plants also take part in the decomposition and absorption of organic matter before it reaches the ground.

In real life things are not as simple and schematic. On one hand, young trees appear in all heights, "masking" the strata. On the other hand, when you make your observation you will surely be in a clear area of the Waterfalls, where the jungle suffered exploitation till the 1930s. Later came different constructions, as buildings for tourism, and more paths.

Also, in the borders of the jungle, where there is more light, the plants grow in palisade, not allowing to see any further. You will appreciate this in the drive to the National Park from Puerto Iguazú town or from the Airport.

"The pristine jungle of Iguazú has let me down" you might say. But there are sectors of the Park where there has been no timber activity, and a good part of the exploited jungle has been recovering its diversity for almost 70 years. Although in many areas which have been very affected in the past, the advance of certain species (such as canes) may be conducing to environments with less diversity.

Change and renewal are constant in this environment. The falling of trees produced by the action of strong winds or by the flooding of the Iguazú River, cause gaps in the forest where colonising species appear.

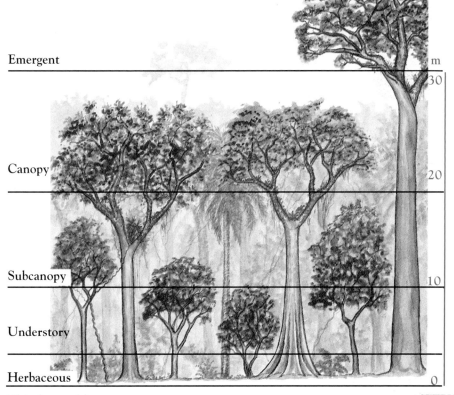

Emergent

Canopy

Subcanopy

Understory

Herbaceous

Thin layer of humus

m

30

20

10

0

The players: Species

In the Iguazú National Park there are known about 200 species of trees, 448 species of birds, 71 species of mammals, 36 species of reptiles, 20 species of amphibians, and amongst the most stunning insects there are about 250 species of butterflies. The jungle this Park protects has the greatest biological diversity in Argentina.

Now then, there is a fundamental question: What is a S-P-E-C-I-E-S?
This is a key concept and can not be omitted in the "Laws of the Jungle".
A simple definition is: A **species** is a group of individuals of common ancestry that closely resemble each other, that can interbreed and leave fertile offspring. The most important characteristic they have in common is their genetic information, which is unique.
All along this book I will mention the word **species** very often and in some paragraphs it might even seem repetitive. It is a concept which has no synonyms.

The Names of the Species

This book is focused on the "rules of the game" and the "positions of several players" in the playing field, without going into detail about the "name and surname of the players". I do not mention the scientific name of the species. I mention the common names of species (e.g. Brown Capuchin Monkey,

Tegu Lizard, Great Kiskadee) or common names which include many species (hummingbirds, toucans, swifts, etc.).

Taxonomy results from conventions established by man to classify Nature. One or more species with characteristics in common are grouped in the same genus; genera in the same Family; families with a lot in common in the same Order; and these in a Class.
In the cases where there is no common name, or if I want to make reference to a group larger than the species, I give a higher taxonomic name, specially genera and families.
For example: Butterflies of the genus *Morpho*, trees of the genus *Cecropia*, butterflies of the family Sphingidae.
Common name: Water Turtle. Its taxonomic classification is:
Class Reptilia, Order Testudinata, Family *Chelidae*, Genus *Phrynops*, Species *Phrynops willamsi*.

Water Turtle

In several opportunities I mention plants and animals of Central and South America or of different continents. It is useful to exemplify concepts and to make comparisons with Iguazú.

The renewal of the players: evolution and extinction

You must have heard about the book *The Origin of Species* by the british naturalist Charles Darwin, published for the first time in 1859. The theory of the evolution of species shown there is probably one of the most important contributions to Natural Sciences.

Of course to try to explain, even briefly, the biological evolution with the rules which it follows exceeds the objective of this small book and I would get into trouble if I tried to do so.

In spite of this, I use the word evolution quite often. For example I mention "co-evolution" during "millions of years" between flowers and their pollinators, between fruits and animals which eat them and disperse their seeds; between predators and preys, etc.

Never-the-less, I think that "co-evolution" and "millions of years" are very difficult to visualise. That is why I am going to give a very coarse example whose only aim is to help to understand key concepts in Nature: **evolution, change, diversity,** and **extinction** of the **species.**

Example: Imagine you are in the perfumery aisle in a supermarket (who can't imagine this example now-a-days? A suggestion: don't buy anything even if you imagine a sale). Dozens of shampoos, deodorants, toothpastes, perfumes, soaps, etc. How many of these products have shown different types and brands during, for example, the last fifty years? Many, without doubt. Let's consider the shampoo (if you are bald you might find it harder to visualise this example, but please try to do so). Most probably when shampoo was invented there was only one type. What do we have now?

One can find shampoo for dry hair, normal hair, greasy hair, hair damaged by the sun, for children, etc. They have appeared by a process of specialisation. New niches have been covered. There are products which have diversified, others have disappeared, some are new, and others are about to appear. There is a lot of competition, but also a great range of co-operation amongst the diversity of products (special offer, if you buy a bottle of our shampoo "Jungle" you can take a bottle of our conditioner "Conserve" for free!).

All these continuous changes result from the co-evolution of consumers-products, although it is probably not that simple.

What forces rule the changes? When and where did each of these products arise? What happens with catastrophes such as hiperinflation? These are some of the questions which we might ask ourselves if we are required to make a simple analysis of the situation.

Let's go back to the jungle of Iguazú to compare it to our stand. We will compare plants and animals with the products of perfumery (apologising to Nature for our lack of consideration in the comparison, but I think it is useful).

A question which pops up is: How many species of different forms of life have originated from each species of bees, hummingbirds, monkeys, orchids, of each of the known species?
The answer is many throughout millions of years. Of course it is not the same to talk about mammals or birds than of insects. The latter is a very ancient group which hardly leaves fossil records and of which we have very little information, even not enough to know how many species there are in the present.
Amongst the mammals it is known that there are 150 species of fossil "elephants", while there are 2 living species. There are about 200 species of fossil "rhinoceros" and there are only 5 living species. There are 250 species of fossil "monkeys" and 200 living species. The Natural History Museums conserve a diversity of bones of extinct species of fauna, which represent a minimum number of those that really existed.

How many species are there in the world?
A good question which has no answer, not even a close approximation.

Studies of the 1980s in the treetops of the tropical forest gave surprising results. For example, one hectare of the jungle in the Tambopata Reserve in Peru can contain up to 41,000 species of insects!

Now-a-days some scientists estimate that for each of the approximately 1,200,000 species that are known, there are about 30 to be discovered in the canopies of the tropical jungles (most of them insects). This would add up to 30 million species in the whole world.
What is known is that there is a direct relationship between the loss of natural areas and the extinction of species. The average duration of a species till its natural extinction varies according to the plant or animal group it belongs to, or the environment it inhabits (earth or water), but in general it is calculated between one and ten million years. There have also been natural catastrophes which have caused mass extinctions, for example that of dinosaurs.

With the actual destruction of natural environments the extinction is occurring a hundred or a thousand times faster. In most cases species become extinct before man can even get to know them.

On the other hand, since the origin of life, the Earth has had climatic changes, drifting of continents, formation of mountain ranges, volcanic eruptions and migration of species. All these processes are still taking place.

In the area of Iguazú, the canyon that originates the waterfalls started forming some 100 million years ago. The environmental conditions of the present days started some 18,000 years ago, after the glaciations that affected the area. Before this, the climate was colder and drier.

Other concepts in examples: The centres of origin and the dispersal of species.
Let's consider the species of mammals in the Iguazú National Park. Are you surprised to find out that the species of cats, brockets, foxes and the Tapir that are nowadays found in Iguazú belong to groups of animals that arrived from North America?
As the Isthmus of Panama was formed, some 3 million of years ago, representatives of these groups migrated to South America. Changes in climate such as the glaciations caused more migrations, many extinctions and the appearance of new species. Amongst them, for example, the Jaguar.

What about the plants?
Iguazú is not a tropical jungle, but there are many groups of species of plants with centre of origin near the line of the Equator. The area of Misiones is the border to their distribution. On the other hand, species such as the Lapacho negro, the Alecrín or the Palmito would have their centre of origin in this area.

Probably the term "evolution" and "Darwin" are associated by many only with the phrases: "man descends from monkeys" and "the survival of the fittest". But evolution is a fascinating and complex topic which started a real revolution in the beliefs of society since the publishing of Darwin's theory, a fundamental starting point of new advances on the subject. Only in the year 1996 did the Catholic Church officially accept the concept of the "evolution of species".

Whoever starts doing some research about this subject will find in each step samples of the overwhelming marvels of Nature. Faith will guide you in the search of the presence and action of God.

Now then, coming back to the examples, they were given because they match the layout of this book: To talk about the jungle without mentioning concepts of evolution would be like playing chess without a queen!

To capture your curiosity, a main objective, the most important ingredients have been placed on the "playing field", and E-V-O-L-U-T-I-O-N is one of them.

To preserve the Rainforest and the Falls is the main goal of the Iguazú National Park.

OURS, is to offer you a nice meal while you enjoy your visit.

RESTAURANT
FORTIN CATARATAS

Parrilla (grill), salads, and other typical meals.

BAR
JAGUAR HOUSE

"*Menú turístico*"; sandwiches, *hamburguesas, lomitos, empanadas.*

Iguazú National Park- Area Cataratas
Ph./fax 03757-491040 - fortincataratas@arnet.com.ar

DRUGSTORE CATARATAS
HOTEL SHERATON

gift-shop, *kiosco*, postcards.

Iguazú National Park
Hotel Sheraton Internacional Iguazú. Ph. 03757-491141

The LAWS of the JUNGLE II

Tell me what you are like and I will tell you where you come from

This heading is too ambitious and can not be fulfilled in real life.

There are many exceptions to the rules of Nature, and even more in the complex world of the jungle. Nevertheless the phrase helps to understand general concepts.

The "design" of the trees in the jungle (see the drawing of the strata on page 17).

• Stratum of emergent and canopy trees: they have predominantly long and straight trunks. The branches grow from the main trunk at a similar height, as in an umbrella. The wide crowns prevail.

• Subcanopy and understory trees: the tendency to look for light determines the crowns of the subcanopy stratum trees to be nearly as long as wide. In the understory, they are longer and less wide.

• Trees growing in the jungle gaps usually have a good leaf coverage, as they receive a lot of sunlight. They are generally coloniser species, of fast growth.

• The texture of the bark of the trees is very variable, although there are many with smooth bark or with a bark that peels off. A possible consequence of this would be to avoid the growth of hanging plants. In species of rough bark or with grooves in them, the growth of hanging plants is facilitated.

• Many trees of the tropical forests have buttresses (like ribs) at their base that help as supports, specially in large emergent trees which usually reach heights of 50 meters or more. In Iguazú the buttresses are less frequent and appear in trees such as the Guapoy (Strangler Fig). The tallest trees of this subtropical forest reach up to 30 to 40 meters.

• The roots of the trees are not very deep: they must obtain their nutrients from the thin, superficial fertile layer. On the other hand, to belong to the jungle provides certain protection from the winds. Although the strongest storms usually causes the falling of the tallest trees, opening gaps in the jungle.

Buttresses

The leaves of the trees in the jungle

The leaves of the trees in the jungle are predominantly oval, with smooth edges and ending in drip tips. This type of shape would help for an easier drainage of the water that falls on the leaf, thus avoiding the growth of algae or lichens on their surface (these would diminish the capacity of capturing sunlight for photosynthesis).

Broad leaves are also very frequently found. This helps in the absorption of sunlight. Those leaves which persist throughout the year are usually thick, coriaceous (leather-like), and waxy.
The trees with compound leaves, such as in family Leguminosae (the most represented family in the Tropical and Subtropical forests in America), do not follow this model. The compound leaves usually belong to trees which need to grow to big heights in a short time.

Some plants which grow in the shade have pigments that give origin to reddish spots in the lower surface of the leaves. The red pigment helps to reflect the light that has passed through the top layers of the leaf and in this way more energy is captured for photosynthesis.

In the Iguazú jungle the leaves are generally smaller than in the tropical forest.
Many species have oval-shaped leaves, with smooth edges and ending in drip tips. The family Leguminosae is well represented.

In your walks you will be able to see the dominant forms of the leaves of trees. The stains on the surface of the leaves generally correspond to lichens and algae.
The use of binoculars is convenient to see the leaves of trees which are out of reach.

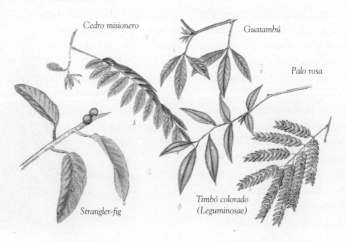

Cedro misionero

Guatambú

Palo rosa

Timbó colorado
(Leguminosae)

Strangler-fig

Seduction of the Flowers

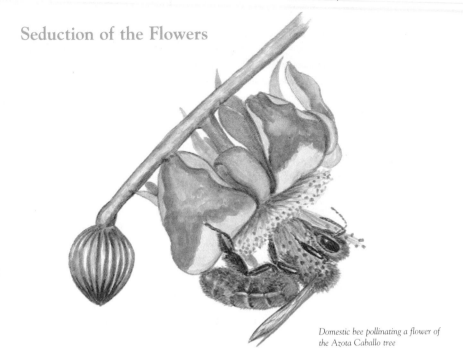

Domestic bee pollinating a flower of the Azota Caballo tree

The climate of Iguazú allows the flowering of plants all year round, although there are certain seasons with more blossoms. Amongst the trees, for example, the Lapacho negro blooms between mid July and mid August and its pink flowers are more attractive before the appearance of the leaves; the yellow flowers of the Ybira-pitá appear in January.

Flowers of a diversity of creepers and lianas bloom at different times, and on the tree trunks flowers of bromeliads, Spanish moss, orchids and many other plants appear.

Now then, what do flowers mean to Nature?

I am not going to go into details about their parts nor their great diversity of shapes and sizes, that range from flowers of a size smaller than one millimetre to the largest flower of the world which weighs seven kilograms of one meter diameter (found in the tropical forest of Borneo and Sumatra). The most important issue is that they contain the organs for sexual reproduction in plants.

The pollen in the male flower must somehow reach the ovule in the female flower so there is fertilisation and formation of seeds.

In many cases both sexes are present in the same flower, but in only a few cases self-fertilisation occurs (see page 61).

In an open environment, as in the Pampas, the main agent of pollination is the wind. Those who suffer from allergies can not ignore the presence of millions of pollen grains floating in the air during spring. This system has been compared to an "air mail" and there are even "long distance" envoys : pollen of the Coihue Southern Beech (*Nothofagus dombeyi*) of the Subantarctic Andean Forest (Patagonia) have been detected in Buenos Aires!

In the jungle there is a basic rule: There are **many species of different plants**, but a **few samples of each**.
Conclusion: the probability that the wind might carry the pollen of one flower to another flower of a different sex of the same species is very low (remember the concept of species).

On the other hand, dense vegetation does not let the wind be very effective, except for some cases which occur amongst the tallest trees. You just need to observe the jungle to realise this.

The result is that in the co-evolution of species of plants and animals a great number of flowers in the jungle have acquired mechanisms to attract different insects, birds and mammals that intervene in the transport of pollen. A very competitive and effective system of a "door-to-door mail" (although in many cases "it rings the bell at the wrong address") that hardly ever rests, as it is only interrupted by heavy rains.

It has been proved that the water can wash away the nectar, destroy the pollen and turn the flower inaccessible for the pollinator.

How do the flowers attract the pollinators?
Fragrances, colours and shapes are the main ways of "seduction". They offer nutritive and sweet nectar or even pollen for food. Most of the animals attracted by the flowers get several parts of their body impregnated with pollen, as you might see in insects.
"I seduce and feed you and you shall carry my pollen to another flower of my species so I can reproduce" are the conditions that evolution has determined without letting the species involved in the process know. It is a complex and sometimes very close relationship of mutual benefit.

The combination of colours, fragrances and shapes of flowers are important in insect pollination. Flowers pollinated by butterflies are generally found in gaps in the forest and are generally red or orange. At least that is the way we can perceive reality. The vision of butterflies, bees and many other insects is sensitive to ultraviolet light. This means they see the colours in a "different way" we do.
The opening of the fragrant whitish flowers attracts the attention of bats and moths (flowers with an awful smell). Those which are pollinated by flies (*Aristolochia*) smell like rotten

meat or dung. A different world appears with darkness. Species of bats that feed on nectar have a very developed olfactory sense and a better sight than their insectivorous relatives, but their sonar system is less efficient. In Iguazú there is no record of this type of bat acting as pollinators, but the moths are abundant.

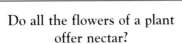

Red and tubular flower of creeper.
Who will pollinate it ?

Do all the flowers of a plant offer nectar?

The answer is no, but let's make some observations.

To produce nectar in all the flowers is more "expensive" that to produce it only in some.

On the other hand, if only a few flowers have a "prize" but without any identification, a hummingbird which reaches the flower for libation (the action of introducing the bill in a flower to feed) will have to visit many flowers in a plant in order to be satisfied. In this way it will leave pollen in a greater amount of flowers.

The species of orchids that offer nectar form the group that contains the least number of flowers with nectar per plant.

Now imagine yourself in the Iguazú forest standing under a great Ybirá-pitá, a common tree, with its thousands of yellow flowers "seducing" pollinators such as bees. On this tree you might see an epiphyte (plant that grows on another plant) in blossom and just imagine it has the pollen ready to be transported. Most probably there are no other plants of the same species nearby with mature female flowers to receive this pollen.

The question is: How does this epiphyte do to compete and to make its pollination be effective if it only has a few flowers (sometimes only one) with which to seduce, and also make the agent it attracts to carry the pollen to a female flower which is not even close? You know by now that it is better to ask the question: What did evolution determine...?, than How does the epiphyte...?

Specialisation is the answer of evolution to this common situation of epiphytes.

In Central and South American

forests, three very special types of "door-to-door" mail used by several pollinators and epiphytes were found (these types of mail also appear in several plants that grow on the ground).

One of these mails is the **"fixed route"** of feeding with **"stops on route"** (trapliners) in certain flowers quite far from each other (there are known cases of plants 15 kilometres apart), all rich in nectar.

Several hummingbirds, bats, big bees and moths (all pollinators that can travel considerable distances) have their circuits with **"stops on route"** in flowers of the families Bromeliaceae, Cactaceae, and many others.

Orchids hardly take part in this method of **"fixed routes"**, but many are specialists in deceiving to attract their pollinating agent.

For example, the flowers of the genus *Dracula* (of the American tropical forest) look like the fungus where the larvae of a species of gnat grow. The female of the insect is deceived and deposits its eggs there, which obviously do not develop, and in the process gets impregnated in pollen.

Sexual tricks are perhaps the most fascinating. There are orchid flowers which imitate the shape and smell of the females of certain bees and wasps. When a male detects them, he is attracted and tries to mate with them!

These tricks were found in a European orchid at the beginning of the century and afterwards in Australian orchids.

In the tropical forests of America it has been proved to happen in the genus *Trigonidium* and in one species of *Oncidium* (a genus found in Iguazú).

The third strategy is that chosen by, better said, determined by co-evolution of the male euglossine bees and more than 600 species of orchids and other families of plants. What is it all about? Flowers produce **"fragrances"** which can be detected by bees at a distance of more than one kilometre. Once in the flower, and with the help of special structures in their legs, bees collect and store the compounds of the fragrances. They afterwards use them to produce their own "perfumes" (chemical messengers for sexual attraction).

Euglossine bees are found mainly in jungles and there are some species in Iguazú. Many feed on pollen and nectar of orchids.

With so much specialisation, there are species of orchids which depend on one species of insect for their pollination. If one of them disappears, the other will follow the same destiny. In the complex, interrelated world of the jungle this might lead to a "trophic cascade" (chain extinction).

To prove any of these relationships escapes our possibilities in one visit to Iguazú. Researchers can spend months and months observing and experimenting to verify them. But if you are lucky enough to observe the fantastic libation of a hummingbird with its long bill in front of a red, tubular flower, or when you find an insect posing on a colourful petal that looks like a marked landing platform, you will know that each detail might be the result of a fantastic co-evolution throughout "thousands and thousands of years"!

If this is how you feel, you are ready to forget the visualisation of the concepts of evolution, change and diversity of the "shampoo" (see page 19) to start visualising the innumerable details of Nature.

An exclusive and fascinating example in Iguazú is the pollination of the Guapoy (see page 52) by a small wasp of the genus *Blastophaga*.
The immature figs (called syconia) of the tree are hollow and inside have hundreds of very small flowers of three types: male flowers, fertile female flowers and sterile female flowers.

The female wasp is captivated by chemical messengers of the plant present in the air. Once it lands on the syconia, it gets in through a small hole and, on its way in, it loses its wings and antennae, thus is deadly injured. But before dying it lays its eggs on sterile flowers, which will be the future source of food for its larvae.

Pollen the wasp carried impregnated on the body from another syconium might remain in the female flowers, ensuring their pollination. When larvae turn into adults, males mate with females and then die.

The new generation of fertilised females leaves the syconium through the hole, getting covered with pollen. It is believed that they have only a few days of life to enter another syconium and lay their eggs, giving continuity to the cycle.

A few days after the females have left, the syconium turns into a mature fruit with very small seeds and will be used as food by many birds and mammals.
The Guapoy is very common in Iguazú, and its syconia, placed close to the axils of the leaves, have a diameter of only 1 cm.

You might see syconia of a larger size (some 6 cm in diameter) in a species of *Ficus* which is originally from India and can also be found in this area. It is a tree with big leaves and there are a few samples planted close to the Waterfalls. It can also be found in some streets in Puerto Iguazú.
Growing directly on its trunks and branches you will see immature (syconia) and mature figs. If you study a sample which has fallen onto the ground you will have an idea of its internal structure.

Co-evolution: the Guapoy and its wasp

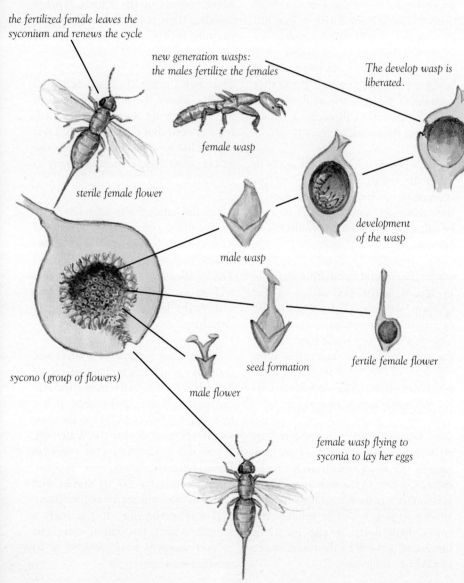

the fertilized female leaves the syconium and renews the cycle

new generation wasps: the males fertilize the females

The develop wasp is liberated.

female wasp

sterile female flower

development of the wasp

male wasp

sycono (group of flowers)

male flower

seed formation

fertile female flower

female wasp flying to syconia to lay her eggs

Adapted from: *An Introduction to Tropical Rainforests*. Whitmore, T.C. 1993. Oxford University Press, New York.

> Nature is a world of competition, but also of co-operative relationships, even at a very high cost for the parties involved.

In Iguazú there isn't a specific wasp to pollinate this tree, so what one usually finds is aborted ovules (it would be the same case as the black spots in bananas, which are not seeds either).

Never-the-less, in the last years it has been observed that some syconia have been fertilised by some species of wasp not identified yet. This implies the risk of the dispersal of seeds of a tree which is not native of the Iguazú jungle. It may start growing in the area, as has happened with other species introduced from other areas (the Uvenia, the Orange tree, the China tree, etc.)

By the way, fruits which are formed directly on the trunk or branches of plants are usually food for fruit-eating bats, as in several *Ficus* (for example the Guapoy), Cocoa plant, Papaya and other plants of the tropical forest. This disposition makes feeding during the bats' flight easier or it also allows them to "park" on the plant as they feed.

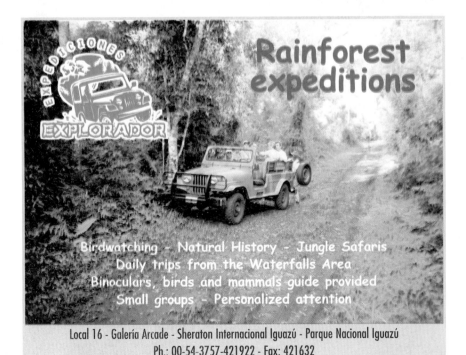

EXPEDICIONES EXPLORADOR

Rainforest expeditions

Birdwatching - Natural History - Jungle Safaris
Daily trips from the Waterfalls Area
Binoculars, birds and mammals guide provided
Small groups - Personalized attention

Local 16 - Galería Arcade - Sheraton Internacional Iguazú - Parque Nacional Iguazú
Ph.: 00-54-3757-421922 - Fax: 421632
explorador.expediciones@rainforestevt.com.ar   www.rainforestevt.com.ar

Temptation of the Fruits

Tactics of seed dispersal

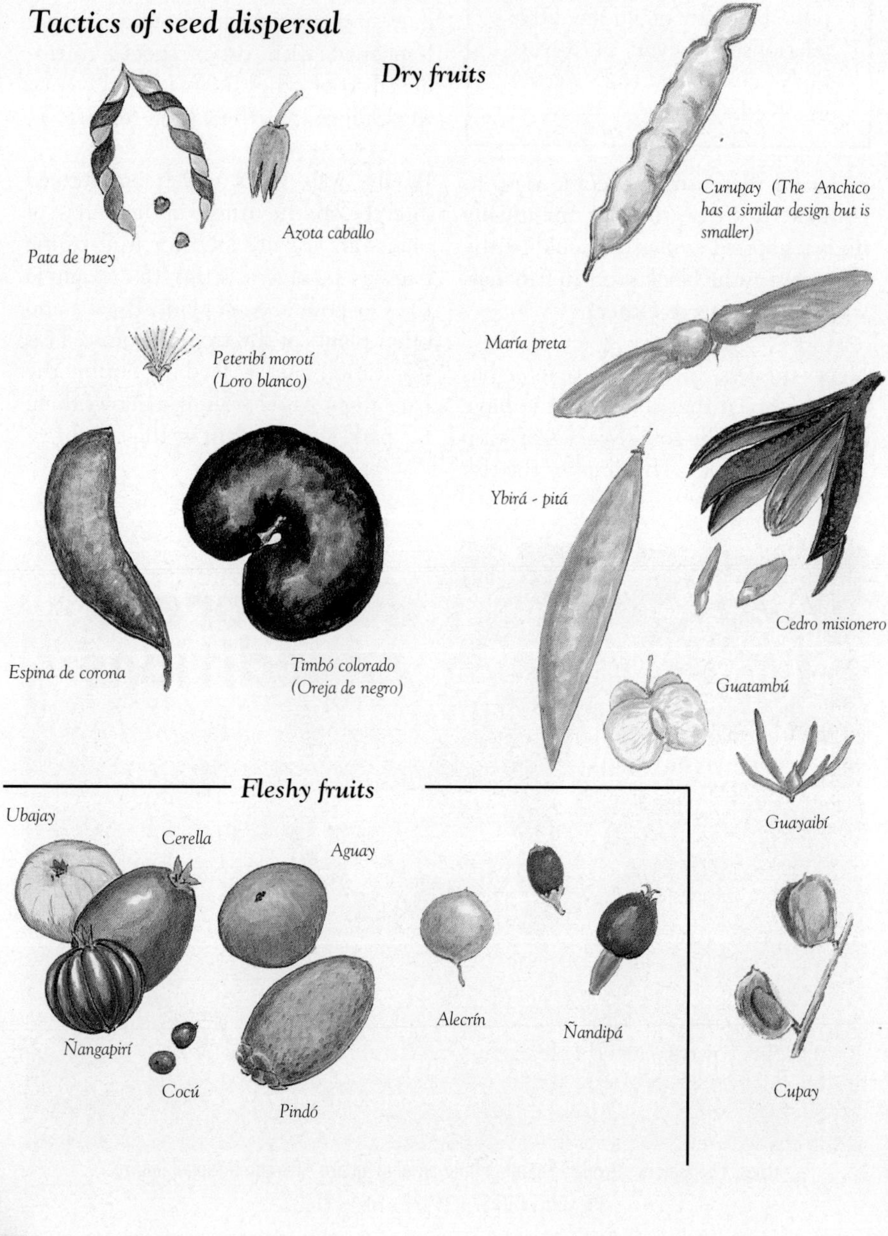

Dry fruits

Pata de buey

Azota caballo

Curupay (The Anchico has a similar design but is smaller)

Peteribí morotí (Loro blanco)

María preta

Ybirá - pitá

Espina de corona

Timbó colorado (Oreja de negro)

Cedro misionero

Guatambú

Guayaibí

Fleshy fruits

Ubajay

Cerella

Aguay

Ñangapirí

Cocú

Pindó

Alecrín

Ñandipá

Cupay

In the jungle, the wind is not the best method to transport pollen or to disperse seeds. Of course there are many exceptions, as is the case of the tallest trees whose treetops are exposed to the gusts of wind and who also have seeds and fruits with adaptations for wind dispersal.

In the Iguazú jungle there are several trees with these characteristics:
The Palo rosa, the Lapacho negro, the Cedro Misionero and the Azota Caballo have small winged seeds.
The Maria Preta has two small seeds joined by their respective wings.
In the Ybirá-pitá and the Incienso tree there are winged pods.
The fruit of the Guatambú has 3 or 4 wings.
The Peteribí and the Guayaibí develop a little fruit on which parts of the flower remain. These open forming a "star" that helps to slow down the fall of the fruit making it twirl as the blades of a helicopter (in this way they can reach a further distance by the action of the wind).
The Orchids have many very small and light seeds, which can be dispersed even by a light breeze.

Other plants have fruit which open violently by changes in the atmospheric humidity. In this way they can send out their seeds at a considerable distance. Examples of this type of dispersal are: Mandioca brava, Pata de Buey, Palo de leche.

The relationship between animals and plants has determined birds and mammals to be the most effective dispersers of seeds in the jungle. Once again diversity is the answer from Nature.

The simplest are the fruits with hooks or bristles that stick to the fur of mammals and are thus transported, as occurs with the Peteribí Morotí.

The fruits which have fallen onto the ground are eaten by rodents such as the Paca, Agouti and Brazilian Squirrel (see page 103), brockets, peccaries, tapirs, and some carnivores of a broader diet such as foxes and weasels. The Timbó Colorado is a gigantic tree (family Leguminosae) whose mature fruit falls to the ground without opening. Its shape is similar to an earflap and has 20 seeds.

To present food in tempting fleshy fruit has resulted in an effective way of dispersal. Diverse shapes, colours and sizes have appeared. Different animals feed on them and later they egest or regurgitate the seeds at different distances from the original plant.
The small fleshy fruit of the Ambay (see page 51) and the Guapoy (see page 52) have a very large number of small seeds. The sweet fruit of the Guapoy can be found almost throughout the year, at different stages of maturity.
The Pindó Palm (see page 54) offers its mature fruit (small yellow-orange round drupes with a sweet seed and

pulp) almost throughout the year.

Small berries (fruits with a thin skin, succulent pulp and many seeds; e.g.: the tomato) are offered by many plants: Alecrín, Cerella, Aguay, laurels (usually dark berries), Yacaratía (orange berries), Guembé (see page 59), caraguatáes (see page 57). The fruits of the Carayá-bola tree open by four valves and have four fleshy and orange seeds.

Many fruits have in their inside a succulent layer covering the seed, called aril. For example it is found in the Cancharana and the Cupay (a dominant tree in the islands of the Upper Iguazú River).

As fruits are maturing, their colours tend to be similar to the foliage and pass unnoticed. Some contain toxins or a disagreeable taste that help them not to be eaten before they are mature.

As they mature they turn red, orange, yellow or other contrasting colours that send the message: **"Food is served!"**

Birds are the main guests to this banquet, although there are several mammals which are also invited, specially the bats, and in Iguazú, the Brown Capuchin Monkey is an important disperser. The dominant red colour of many fruit can represent not only food, but also other resources such as carotenoids with Vitamin A, important for the sense of sight and as precursors of different pigments which give colour to the feathers in birds.

The fellow diners in the jungle have "menus based on fruits" throughout the four seasons. Although there might be times of scarcity, they are most abundant in spring and summer. This is one of the reasons why the births of many mammals that have fruit as part of their diet occurs during these seasons.

It has been observed that amongst the birds, in groups as the guans (in Iguazú, the Black-fronted Piping Guan and the Rusty-margined Guan), toucans, and trogons often swallow the fruit directly and would be hardly sensitive to their taste. They egest or regurgitate the seeds in different places according to where they travel.

Instead, the colourful and small fruit-eaters (see page 92) prefer the sweet tastes and peck and eat slowly. There are seeds which fall onto the ground and others that are ingested and later defecated. Many seeds are hard and are not digested and the seeds of some species can remain latent till the environmental conditions are optimum for their development. But as time goes by, they will be predated and destroyed and only a few will germinate to produce a new plant.

Many parrots destroy the seeds with their potent beaks, and some even digest them, while the pigeons degrade them with their digestive juices. This is why, generally speaking, they do not act as dispersers.

The abundance of food on trees and the fact they give protection against predators, favoured the evolution of mammals adapted to live on trees. Among them, the Brown Capuchin Monkey (see page 102) added several fruits to its broad diet.

In areas with shade, as in the Macuco path, you can find the bushes known as Pariparoba (genus *Piper*, group to which the Black Pepper also belongs). They can be identified easily by their erect infrutescence (group of fruits) which looks like a small candle. It has very small fruit which can be captured in flight by fruit-eating bats (of the genera *Carollia* and *Sturnira*, see page 105) which transport it to it eat in a

safe place. After going through the animal's digestive system, some seeds fall on appropriate ground and originate a new plant.

Walking along the Iguazú paths it is possible to observe several disperser birds (see page 89). On the ground you will be able to find seeds and fruits, some healthy, many pecked, and maybe some with fungi or insects feeding on them.
Consulting the "Seasonal Fruits" Chart (see page 117) and the Maps with the location of flora in the Circuits (see page 120) you will be able to find samples which represent different methods of dispersion "designed" by Nature.

When lunch or dinner time arrives, you can choose among the varied fruit which are served in Iguazú. I am sure you will not act as a seed disperser, but do not feel like an uninvited guest.

Apart from the fruits, the plants provide a great diversity of food products. The agronomic studies have allowed the creation of more productive hybrids and better adapted to different climatic conditions. But we have only used a few species, and the devastation of the forests and other environments can be extinguishing resources of unsuspected value.

A Matter of Size

A small or middle sized fruit has more chances of being eaten. This increases the possibility of dispersion of its seeds, but also its predation. This is a strategy used by many plants which grow quickly, as in the coloniser species.
On the other hand, the big fruits hold more reserves, and their seeds have a greater possibility of developing, but there are less animals who disperse them.

Plants Growing on Other Plants: Epiphytes

Caraguatá

The name epiphytes has been given to the plants that grow on other plants without taking nutrients from their support, and with the aim of obtaining sunlight for photosynthesis. They are not parasites of their support. A clear proof is the fact that they can develop on the rocks, as you can see in Iguazú.

Their diversity is so large that it is considered that of the nearly 250,000 species of flowering plants known worldwide, 10% are epiphytes. They range from simple algae, lichens and mosses to ferns and flowering plants such as orchids, bromeliads (Spanish moss, caraguatáes), cacti, among many others.

As they grow on trees (apart from cap-turing more sunlight), the wind acts in a more effective way for dispersal of their pollen and seeds; and if they depend on animals, they offer an easier access to many pollinators and fruit dis-persers.

The great disadvantage is that, as their roots are fixed to the bark of trees, they don't have access to the nutrients and water from the soil. Growing on very tall trees they will be more affected by drought and frost, while in the lower levels the conditions are more stable, although there is less light.

Some help is obtained by the different texture of the bark and from places such as the axils of the big branches.

Let's take for example the Pindó Palm. Its straight trunk and smooth bark is not the best terrain to grow and to sur-vive as a flowering hanging plant. Most probably the lichens will be present, specially if the Pindó grows in the jun-gle (there is more shade and less evap-oration) and not in a gap. This is the case of noticeable pink lichens. Many of these plants can obtain nitrogen (one of the main nutrients) directly from the atmosphere or from rain water.

Many trees lose their bark in strips or sheets and with them they lose the epi-phytes, while others have chemicals which do not allow the growth of plants on them.

On the other hand, the trees whose bark is rough and spongy and does not fall off can retain some water and nutrients and offer good possibilities for the germination of epiphyte seeds. The Coral trees (*Erythrina sp.*) for example, have barks with deep grooves where abundant hanging gardens are formed. Other trees that have a rough bark are: Palo Rosa, Lapacho negro, Peteribí and Azota Caballo.

Dead leaves, sticks and other particles are added to the substrate where the roots of the epiphytes develop. In some cases there are ants' nests which also give nutrients.

The lack of water is a risk which is always latent for a plant without roots in the soil. That is why it is not surprising to find Cacti (for example, from genus *Rhipsalis*), a family of plants we associate with dry environments, amongst the epiphytes of the damp jungle of Iguazú.

Are there Cacti growing on the ground in Iguazú? There are not many, but on dry and rocky soils there can be samples of the genera *Cereus* and *Opuntia*; in the Lower Circuit some lying Cacti are found; and the *Brasilocactus sp.* is a columnar cactus which grows on the sunny banks of the Lower Iguazú River.

The spray of the waterfalls keeps a high environmental humidity favouring the growth of epiphytes: The caraguatáes (see page 57) with its colourful inflorescences; the Güembé (see page 59), of enormous leaves; orchids (see page 60); spineless cacti; ferns, lichens and mosses.

The great abundance of epiphytes can damage the trees they hang on to. The excessive weight on their branches could cause them to break, or the dense covering can compete in the capturing of light for photosynthesis.

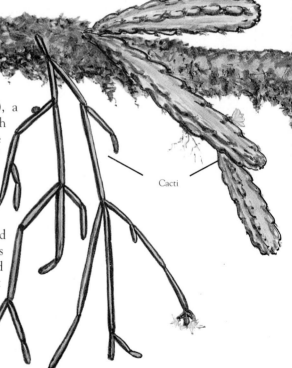

Cacti

The hemiparasites do affect the tree in a clear way, as they obtain sap from their support with a type of absorption "hose" (haustorium). They do have leaves with chlorophyll.

In Iguazú, the Caá-votyrey or Yerba del Pajarito is a hemiparasite which can be better seen in winter on the trees that lose their leaves. The hummingbirds help in their pollination and their little fruits are appreciated by the Euphonias. The defecated seeds fall covered by a sticky substance which allows them to be fixed on to some plant, as they do not develop on the ground.

Lianas

Epiphytes are very abundant in the tropical and subtropical forests (specially in America) and are well represented in Iguazú.

But they are also abundant in Patagonian forests, with a high diversity in the areas of greater humidity corresponding to the Valdivian forest. There are predominantly mosses, lichens and ferns growing on trees. Large lianas, instead, are woody creepers more common in tropical forests, well represented in the Subtropical Forests of Argentina. What restricts the growth of big lianas in the South or in areas where the climate is dry?

Part of the answer is in the long shapes which allow the exchange of heat. Lianas have a very long stem that goes twisting to climb on the trunks and they also count with the help of tendrils (a kind of hook) and other means of fixation.

As they reach the treetop they can become hanging, and they generally keep only the leaves that receive light, up high. But if they are very exposed to the sun, the high evaporation can desiccate them. On the other hand, with low temperatures their sap can freeze.

Another inconvenient could be compared to what happens with a watering hose pipe. The transport of liquids requires a high pressure, and if it twists, circulation could be difficult or even cut off. Of course, the mechanisms of transport are more complex in a liana than the simple running of water in a hose.

Their spiral growth or in steps to climb tree trunks would give them more resistance to twisting. From this type of growth derives the name of "escalera de mono" (monkey's ladder), that can be seen in Iguazú.

Once the liana is on the tree it helps other lianas to creep on it. In a short time, they will extend on to neighbouring trees forming a complex net. When a tree falls by the action of the wind, the lianas can cause the same to happen to the neighbouring trees. This will open a gap in the jungle initiating a new cycle with the advance of colonising species. Seeds which might be latent, such as from the Ambay, will have their opportunity of forming a new plant.

The San Martín Island is the Waterfalls area where the largest trees are found (as it has suffered less exploitation in the past). The paths of the island are a good place to observe lianas.

Try to imagine how complicated the transport of sap is in these climbing plants as they can be several meters long. Try finding trees covered with lianas to see how they intertwine with neighbouring trees; observe if they have epiphytes growing on them. Discover the "monkey's ladder" liana or note the spiral growth of others. Observe if there are leaves or flowers at different heights on the lianas, or if they are more abundant in the highest part where there is more light.

Asking questions and trying to find the answers is part of the "game".

Creepers: lianas and vines

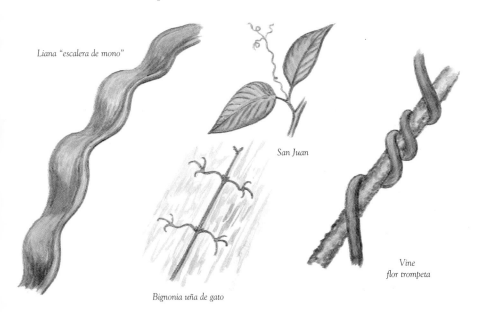

Liana "escalera de mono"

San Juan

Bignonia uña de gato

Vine
flor trompeta

The Hidden Laboratory

We all know about the prickly thorns in the stems of thistles or the little thorns in the fruits of the tunas. They are a very good protection against herbivores.

Many plants have a mechanism of defence which is more subtle and that we can not detect by naked eye: chemical substances.

It might surprise you to know that the *Dieffenbachia* (family Araceae, see page 59), known as "tropic", an indoors ornamental plant frequently used in many countries, has oxalic acid in its sap which can produce severe irritations in the mouth.

Do you know which toxin can be obtained from the bitter almonds, pits of plums and cherries or from the edible tubercles of manioc? Nothing else but cyanide! Its concentration in some of the varieties of manioc can be dangerous for man, that is why it is important to peel them carefully, clean and boil them well before eating them.

Leaves of Carayá Bola attacked by herbivores

Leaves with chemical defences?

Alkaloids (such as morphine, cocaine, caffeine, nicotine), saponins, cyanogenic glycosides (from which cyanide is a derivative), terpenoids and rotenone form part of the chemical arsenal of plants. More than 15,000 chemical products have been isolated from plants coming from different environments, but the greatest diversity belongs to the jungles. They are called secondary compounds, although the function of many of them as a mechanism of defence is still argued or unknown.

Why has such a great diversity appeared? Has it been the action of evolution, in this case in the relationship between the one that eats and the one that is eaten? You might already have an idea of the answer, although the defences are not only against herbivores, but also against bacteria and fungi.

In mammals, alkaloids can interfere in the functions of the liver, in the production of milk, and can even cause abortions or congenital defects. Herbivores would reject plants that contain them because of their bitter taste.

Several alkaloids produce hallucinations in man. Aborigine tribes use them in their rituals, while in modern society a lot of people destroy themselves with the use of the derived drugs.

Nicotine can be an effective insecticide in gardening, as all those who have taken care of plants might know; and we all have an idea of the effect it has on smokers. Terpenoids are harmful for the fungi which are cultivated by Leaf-cutter ants (see page 69).
Latex is not considered a secondary compound, but it presents defensive functions. It appears in several families, such as the Moraceae (includes berries), among which you can find the Guapoy and Ambay. On one hand it turns sap into a dense substance which is hardly acceptable for insects with sucking mouthparts or for Leaf-cutter ants, which get their mandibles "stuck" while cutting the leaves. On the other hand, it would act specially in the healing of wounds and in avoiding the entrance of micro-organisms.

It has been proved in several tropical trees, which are the favourite of herbivores, that young leaves contain a low concentration of toxins. But in a species of howler monkey of Panama, it was determined that its choice of leaves is based on their hardness, fibre content and nutritive value rather than their toxicity.
There are six species of howler monkeys in the jungles or forests in Central and South America. The Black Howler Monkey is found in the east of Chaco and Formosa, north of Santa Fe, areas of Corrientes and is scarce in Misiones. The Red Howler Monkey is very scarce and is generally found in forests of Paraná Pine trees (*Araucaria angustifolia*) in the Northeast of Misiones.
The Tapir feeds on several species of plants and its proboscis, which is very sensitive to smell, would partially help it to detect harmful toxins.

Insects are the main targets of these chemicals and they are constantly fighting a battle against plants, where some attack and others defend themselves. Several plants produce substances which are similar to juvenile insect hormones, which stop the growth and the shedding of the exoskeleton (an example is the hard "carapace" of the beetles) of these animals.
Many animals have specialised in inactivating certain toxins and thus feed on a determined group of plants. For example the caterpillars of the *Heliconius* butterflies (see page 65) feed almost exclusively on plants of the genus *Passiflora*, which contain substances from which cyanide derives.

Plants in a natural environment have different chemical barriers and because there is certain diversity, they can control the existence of insects which are plagues. But our "monocultures" have given "freewheel" to several insects. A field of wheat, for example, does not present any natural chemical barriers nor predator animals, and at the same time it offers an over-abundance of food.

That is why, regardless all the controls made by man, it is estimated that the plagues of insects cause a yield loss greater than 15% in the whole world each year.

that generally where these plants grow the concentration of herbivorous insects is greater.

The chemical substances these living laboratories (the plants) make are used by the inhabitants of the jungles (who know many secrets) for different purposes. The Guaraníes were the human group which dominated in Misiones before the civilisation by the white man. Many of their cultural values and uses and knowledge of Nature have been conserved. The wise man Moisés Bertoni spent years of research in Paraguay during the last decades of the XIX century and beginning of the XX century. He was one of the first to rescue and record the knowledge which was transmitted orally from generation to generation.

Which plants protect themselves with chemical weapons?

Ferns (see page 56) have little diversity of toxins. Among the families of flowering plants, the distribution is not even. It is also known that it can vary according to the individuals, depending on factors such as the soil. For example: In soils poor in nitrogen the plants can not synthesise alkaloids; the manioc presents a low concentration of toxins in fertile soils, while its concentration is very high in poor soils.

The colonising species which must destine their energy in growing very quickly, usually do not produce toxins. As it was expected, it has been proved

Nowadays, plants are raw matter for the elaboration of about 45% of the pharmaceutical products of the world. However, only 1% of the species in the jungles have been studied for this purpose. They might contain the cure for many diseases.

Güembé

Some plants in Iguazú with a "chemical arsenal"

- Alecrín: leaves and seeds with cyanogenic glycosides (from which cyanide is obtained).
- Timbó Colorado: leaves, bark and fruit with saponins.
- Sangre de Drago (genus *Croton*, includes well known ornamental plants): it receives this name due to the red exudation of resin, tannins, gum and sugar it liberates when the bark is damaged. Its seeds are considered toxic.
- Ceibo: seeds with several alkaloids.
- Yerba Mate: leaves with tanoids and caffeine (matein).
- María Preta : with saponins (as in all members of the family Sapindaceae to which it belongs).
- Rabo: with rotenone in its fruit.
- Güembé: leaves with calcium oxalate (common in the family Araceae).

The widest variety of books about PATAGONIA

Also souvenirs, handicrafts, postcards, maps and T-shirts

GALERÍAS PACÍFICO, BUENOS AIRES
Florida 733 PB Local 227, phone 5555-5267
e-mail: pazosmario@hotmail.com   www.patagoniasur-ar.com.ar

Colour Signals

The *Morpho* butterflies, of shiny blue metallic colours, are one of the most beautiful and colourful manifestations of life in the Waterfalls Circuits. One species has a wing span of 15 cm. The *Heliconius*, on the other hand, attract us with their stunning combination of reddish orange with black, yellow or white.

If the jungle is a world full of predators, why should butterflies be so attractive? It seems like an open invitation to be eaten. Both groups mentioned before have species distributed in Central and South America. Naturalists of the last century, such as Henry Bates during his trips in the Amazon and Thomas Belt in his experiences in Nicaragua, made observations about the meaning of the colours. They might represent signs of warning, a new example of co-evolution (see page 19), in this case in the relationship predator-prey.

The *Heliconius* butterflies feed almost exclusively on pollen and nectar of the *Passiflora* flowers (genus of plants to which the Passion flower creepers belong). These plants produce substances from which cyanide derives, to protect themselves against herbivorous insects. But the *Heliconius* are immune to these substances. The caterpillars ingest it as they feed on leaves and they synthesise their own poison, which is also present in the developed butterfly.

The predator with a capacity of learning (as the birds) which captures one of these butterflies will know it must not try it again. It will remember the colours of the *Heliconius* which act as a great luminous signal warning **"do not try to eat me, I am toxic"**.

There are other groups of toxic butterflies that also show warning colours similar to *Heliconius*. As it is the "dangerous clothing" which is most abundant, there will be a greater

Heliconius butterfly

number of predators which will know the message.

The combination of colours which contrast with the environment, such as red, black, yellow and orange, appear as a warning sign in several groups of animals. Amongst insects it is common in wasps and bees, in the caterpillars of butterflies and many others. A good example amongst the vertebrates in Iguazú would be the poisonous coral snake, with its red, black, whitish or yellowish rings.

The innate aversion (not learned) towards corals has been proved in two species of birds (one of them is the Great Kiskadee, see page 86). The lethal poison of this snake does not give a second chance to the victim.

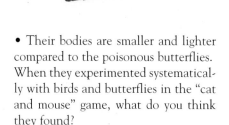

Morpho butterfly

There are many species which mimic dangerous species that live in the same area but without themselves having any type of mechanism of defence. For example, there are butterflies whose colour pattern resembles that of the *Heliconius*, without being toxic themselves- like the *Marpesia sp.* found in Misiones (of the family Nimphalidae)- and the Banded Hognose Snake resembles the pattern of colours of the real corals. When a predator meets them he might be deceived and will not attack.

Why should some butterflies which are not poisonous (and not imitators of toxic species), such as the *Morpho*, be so colourful?

In search for an answer to this question, some biologists studied hundreds of individuals of more than 120 species of butterflies with different levels of toxicity.

The results determined:

• The non poisonous butterflies possess a great mass of flight muscles and larger wings.

• Their bodies are smaller and lighter compared to the poisonous butterflies. When they experimented systematically with birds and butterflies in the "cat and mouse" game, what do you think they found?

As they expected, the non poisonous butterflies were agile fliers capable of making swift moves and thus evading a great number of attacks made by the birds.

The *Morpho* seem slow and easy to catch in normal flight. But in a dangerous situation the aggressor must be very quick in order not to fail. Its colours would mean: **"I accelerate too fast, you will not be able to catch me!"**.

Instead, when they are resting with their wings folded, they show a camouflage colouring (see page 47).

The colour in the plumage of birds is also very important. They can intervene in the attraction of their mate , in the courting of the female and in the delimitation of territories. This is the

Apart from warning and cryptic colours there are other signal codes in the butterflies' wings that neither their predators or us can distinguish. The sight of these insects is sensitive to ultraviolet light, the design and colours they see in the wings establish some type of communication between them, for example the one of sexual encounter.

case of the males in many species of hummingbirds (see page 87). Apart from the colour of the pigmentation of their wings there are metallic shines which result from optical effects.

The manakins are a group of very territorial birds with complex courting rituals where the colour of the plumage of the males is very important. Their main food are small fruits and as these are abundant they must not spent much time and energy in their search.

What do they do with their time?
For example, in the case of the White Bearded Manakin (very territorial; of low density in Iguazú) it has been proved that during the mating period the males spend 90% of the day competing for the female in an elaborate courting ritual (lek polyginy).
The number of cases of "visual signs" are uncountable in the Animal Kingdom. Apart from the colours, the sounds are also fundamental. Specially in the jungle, where visibility is reduced, mainly during the night.
In birds, the sounds can play an important role in delimiting territories, in the attraction of the couple or as signs of alarm. It helps us to identify them and they also charm us with their "music".

But as we are "blind" to the ultraviolet visual spectrum, we are also "deaf" to the messages in ultrasound frequency. Let's see an example: The moths are predated by the insectivorous bats. To avoid being detected, some of these moths interfere or disperse the ultrasound waves they receive from their enemies. Others posses toxins and are distasteful; the ultrasound waves they send back to the bat warn (as does the colour of *Heliconius* to the birds): **"Do not eat me, I am toxic!"**.

You might find some answers to your questions as you read this book, but it is also possible that many more questions will arise as you advance in your observations of Nature. This is a good sign.

Hide-and-seek and Masquerades

To show off or to pass unnoticed. These are antagonic options in Nature, which can be modified by many variables. It can be as quick as the simple change in position of an individual, or according to the stage of life it is going through.

What do the stains in the Jaguar, the bands in the fur of the Giant Anteater or the mottled fur of the Paca mean? Well, they are something similar to the bands in the Zebra. A camouflage which tries to hide the contour of the animal in its environment, either to pass unnoticed to predators or to be able to stealthily get close to a prey, as in the case of the Jaguar.

To remain static and merge with the background is an effective way to camouflage. Insects are masters in this art, and in the jungle there are fantastic examples of this. Many blend in with several objects such as the bark of trees or lichens (moths); small twigs or sticks (the stick insects, insects of night activity; the caterpillars of the Geometridae moths); leaves (Pieridae butterflies) and other details of Nature. The butterflies can attract a lot of attention while flying with their wings

Geometridae caterpillar

open. But while at rest, in many of them, the wings remain in a vertical position and they frequently show in their underwings a camouflage or cryptic colouring.

Ocelli on wings?

In *Morpho* and in other groups, they also present small ocelli (drawings which resemble eyes). Experimenting with predators it has been proved that, when being attacked, the ocelli attract the aggressor, giving the butterfly greater chance of escaping undamaged.

Other groups of butterflies (in Iguazú the *Caligo sp.*, of the family Brassolidae, with dawn activity) and even some frogs have large ocelli they keep hidden and only show when a predator arrives. This mechanism would act by frightening or immobilising the attacker giving enough time for the prey to escape.

Yacutinga Lodge

Suatainable Tourism at Yacutinga Lodge & Wildlife Nature Reserve

Yacutinga Lodge has been created to offer very comfortable lodging to Nature lovers who would like to discover the Argentine Subtropical Rainforest. Its careful design in harmony with nature offers various possibilities to participate on organized walks and excursions in the Reserve area.

These walks are based fundamentally on observations and interpretations of wild fauna and flora, therefore inside the property nine perfectly signed path have been designed, which transit through the various natural areas.

Additional to these paths are the navigation on the upper Iguazú river, floating trips on the Riacho San Francisco, various observations towers and the spectacular cat-walk of nearly 80 meters length and 6 meters height. Expert local English speaking guides with biological knowledge conduct all these excursions.

Worth mentioning is the richness in flora and fauna present in these 560 hectares. The property's surface conforms a whole Peninsula with 6 km of river border to the Iguazú River.

The property has been declared Private Natural Reserve through a Private Refuge program promoted by the *"Fundación Vida Silvestre"* (representing the WWF in Argentina) since species close to extinction can be found.

FUNDACIÓN VIDA SILVESTRE ARGENTINA

The Lodge buildings follow smoothly with the natural step of the terrain and maintain harmony with the surrounding jungle. Only 4 hectares of the 570 total have been used for the facilities of the lodge and special care has been taken to use local materials, such as stone and large wood pieces gathered from naturally fallen trees that have been incorporated into the general architectural concept.

An absolute minimum of forest has been disturbed to accommodate the buildings. The main building, which contains the reception, an indoor restaurant and a terrace bar, provides a special atmosphere enhanced by the Palmetto palm forest formation, which surrounds the whole place. The facilities consist of separate sleeping areas called "Butterflies", spread through the Jungle around the common areas.

Each Butterfly contains four private suites, all with the comfort of 42 sq. meters. The suites provide accommodation with rustic details comprised of an entrance porch, the bedroom, leading up to the bathroom. Each suite has private facilities with 24-hour warm water.

Staying at Yacutinga Lodge means entering in communion with Nature

Reservation Agent in Iguazu
AGUAS GRANDES EVT
Mariano Moreno 58
Ph 54-3757-421140 - fax 423096
aguasgrandes@aguasgrandes.com
www.aguasgrandes.com

Reservation Agent in Buenos Aires
SAFARIS NATURALISTAS EVT
25 de Mayo 758 - Piso 10 «G»
Ph/fax: 54-11-4312-0928/4313-5155
safaris@arnet.com.ar
www.safarisnaturalistas.com

For more information, please visit www.yacutinga.com or e-mail: *yacutinga@yacutinga.net*

Walking slowly and observing carefully the surface of leaves, stems and trunks, you will be able to detect some interesting things.

If you carry binoculars and you find a male Surucua Trogon (see page 92) you will be observing one of the colourful birds of the Iguazú jungle. Nevertheless, the static and quiet posture it keeps on the branches and the lights and shades of the jungle determine that very often one passes beside one of these birds without noticing. This happens with many other species.

A continuous game of hide-and-seek and masquerades is lived every day in the jungle and only some of these mechanisms have been discovered.

We also give more importance to the visual signals, such as colours and shapes. But there is also a world of sounds, smells and other senses, including vibrations, which are difficult to detect and interpret.

Anyway, there are enormous amounts of details which you can look out for. Remember: "eyes wide open" and "ears on end".
Small things can hold surprises.

Flora

Ambay

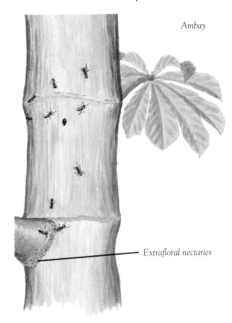

Ambay

Extrafloral nectaries

The Ambay is one of the nearly 50 species of the genus *Cecropia* found in Central and South America. It belongs to the family Moraceae, as does the Guapoy.

Cecropia are one of the first trees which grow in the gaps which appear in the jungle or in the banks of streams; and - after the action of man-, in deforested areas, or at the sides of roads. They are coloniser species or pioneers.

In the area of the Waterfalls, in the Iguazú National Park, the Ambay is one of the most common trees.

They have a straight trunk with hollow compartments separated by partitions. They branch up high to form a wide treetop and they posses very few plants growing on their trunk (lichens dominate). They do not spend energy nor raw matter to synthesise toxins against herbivores (they have latex, a characteristic of this Family).

Their large and parasol-like leaves need to be exposed in gaps with good light. The fertilised flowers produce an impressive number of seeds (in some species up to 900,000 seeds per plant in each fruit), giving small fruits quite frequently.

Don't these characteristics show us some type of strategy?

It seems that all the energy of the tree is used to obtain a fast growth and dispersion, typical of the coloniser species.

A sample of *Cecropia* of a tropical species grew five meters high in only one year. Yes, five meters in only one year!

Another noticeable feature of the *Cecropia* is that Aztec Ants (genus

Azteca) live in its hollow trunk. These aggressive ants defend their territory, in this case the tree, against several herbivores, for example Leaf-cutter Ants. In exchange they obtain refuge and food from the extrafloral nectaries, a type of "gland" which produces nectar, found in the axils.

A great concentration of these ants can represent a disadvantage as it means a tempting banquet for predators such as woodpeckers (see page 84). It is also possible that ants are not such an efficient means of defence as would be the chemical toxins produced by other plants. This is the reason why in some *Cecropia* the leaves can be affected by herbivores.

If you stop in front a sample of Ambay you may see it for yourself. The extrafloral nectaries and the small holes through which ants enter the hollow trunk are visible. If the tree has fruits and you can hide somewhere close, with a little patience you might observe different birds, such as Plush-crested Jays, tanagers, euphonias, or even a toucan, in search of food.

To be alert and have patience are two important requirements to observe and feel Nature.

●●●

Guapoy (Strangler Fig)

Guapoy: "kids grow as time goes by…"

Many people know the ornamental *Ficus* which is found decorating offices, homes or gardens. In the world there are about 900 species of this genus of the family Moraceae. One of their common features is the production of latex.

In Central and South America and also in Africa there are several *Ficus* which have gained the ill-fated fame of "stranglers". Amongst them the Guapoy a tree which is frequently found in Iguazú.

Let's start describing the life cycle since its small fruits are eaten by several birds and mammals. The seeds will be defecated or regurgitated and might produce a plant if they fall in an appropriate place.

The axil or the breaking of a branch of a tree where there is some humus is an appropriate place for a beginning. The Guapoy will start growing as a small plant of short roots and a shoot with leaves. Later the roots will start growing downwards, wrapping and "strangling" the host tree. As it reaches the ground it will be connected to a great source of nutrients and water.

With time, the host tree that has remained locked in will "exhale" (in most cases). If we had to write a "post-mortem certificate" most surely the forensic doctors would not coincide in their diagnose. It would be a combination of "very slow death by strangling" (its circulation in the vascular system would be obstructed) as the tree can not grow in width, or "death by starvation" (as photosynthesis can not be done properly as its leaves are covered by those of the invader). On trees where an unhealed wound was a good substrate for the seed of Guapoy, it could be "death by rotting while still alive", as the decomposition advances from the inside and at the same time provides humus for the new plant.

It also depends on which tree is the host. Palm trees hardly grow in width and in Central America there are cases where both trees have developed, one inside the other, and this could also be seen with other species in the Iguazú jungle. In the Waterfalls Circuits you can find Guapoy which have attacked the host trees in different degrees. It has to be mentioned that in the tall forests of Iguazú, samples of Guapoy of a great size (more than 20 m high) can be found growing directly from the ground, without the need of "strangling" any tree.

Another amazing characteristic of these *Ficus* is their specialised pollination (see page 29) which allows them to have samples with fruit all year round. Wasps in charge of their pollination are born, feed and develop inside the future fruit and only have a few days of free life.

The Guapoys are not synchronised by atmospheric conditions. As it has been said, they "grow their own pollinators".

• • •

Palm Trees: Pindó and Palmito Edible Palm

Pindó Palm

In subtropical forests -as in Iguazú- a low palm diversity is characteristic, different to tropical forests of America where there is a great diversity.

The Pindó is the dominant species and it adapts to different atmospheric conditions. In the Waterfalls area you will see isolated samples which can be 20 m tall in the gaps and taller samples in the jungle, as they need to compete to capture more sunlight.
It has a fast growing trunk, although it is not considered part of the first colonising species, as is the Ambay.
Palm trees hardly grow in width, that is why, if the trunk is damaged, the scar will remain for its lifetime. Usually no flowering plants grow on the smooth bark of the Pindó (except for areas where the spray of the waterfalls provides continuous humidity). Because of their type of development, the lianas find it difficult to grow on them.

On the other hand, the roots of palm trees do not branch much and the new ones grow from the base of the stem and can appear on the surface, as you will be able to see in large samples.
Their large leaves fall and are replaced and the small flowers with separated sexes join forming inflorescences (group of flowers). The blooming of the Pindó lasts for some time, it is pollinated by insects and produces small sweet fruit almost all year round. These are food for birds such as toucans, parrots, jays, the Rusty-margined Guan, which swallow them whole; for mammals such as the Brown Capuchin Monkey, Coati, Agouti, Brazilian Squirrel, peccaries and even the Tapir; and for reptiles such as the Tegu Lizard. The Red-rumped Cacique (see page 94) uses the fibres of the Pindó to build its nest. Their decomposed trunks which are still standing are a good place for the building of parrots and toucans' nests, and also for the development of insects' larvae, which can be food for woodpeckers. The Guaraní aborigines fry the larvae of wood beetles in fat to eat them.

The Palmito Edible Palm is a species difficult to find in the Waterfalls area, but can be found in other regions. It has been intensely exploited because of its "palmhearts", whose extraction means the death of the tree.

This palm is protected by the Urugua-í Provincial Park, the Iguazú National Park and the neighbouring Do Iguaçu National Park, in Brazil.

As they have fruit during the season with less food (winter), the Pindó and the Palmito edible palm are very important for the feeding of many birds and mammals of Iguazú.

• • •

Bamboos

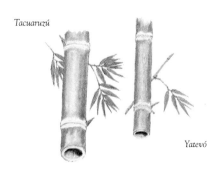

Tacuaruzú

Yatevó

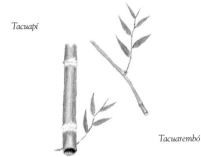

Tacuapí

Tacuarembó

The abundance of bamboos is characteristic of the Iguazú jungle. These plants grow in a vegetative form for a period of time, which varies according to the species, which can range between only one year to about 120 years.

Do you like fishing rods?

I'm not keen on fishing, but I have a faint idea of the great flexibility of the artificial materials used to make modern fishing rods. In the early days they were made of bamboos. These plants are light and many are hollow, and they have in the outer part of their stem long, elastic fibres which are very resistant.

This is the ideal structure for plants with shallow roots and tall stems as the Tacuaruzú, found in Iguazú, which can be 30 m tall.

In some Asian countries, the bamboos are a symbol of "fight against adversity", as only very strong winds can knock them down and if this occurs they can still recover.

In their last stage of life, the bamboos produce a mass flowering, they are pollinated by the wind (see page 26), they produce seeds and die. Little is known about the factors that start the flowering, although it is known that in some species it is determined in its genes.

In Iguazú there are four species of bamboos. The largest is the Tacuarazú which flowers every 25 to 30 years (the last flowering occurred in late 1970s). It usually grows on the banks and islands of big rivers as the Paraná and Iguazú Rivers. This species is highly resistant to fungi and insects.

The Yatevó is a hollow bamboo which can be 15 m tall; its stem has thorns, a protection against herbivores. The Tacuapí can be 10 m tall and is also hollow, but its surface is rougher and has no thorns; it is an invader species in deforested areas. Lastly, the Tacuarembó bamboo-with a solid and thin stem- belongs to the genus *Chusquea* , which includes 90 species distributed between Mexico and the south of Argentina and Chile. The southernmost species of *Chusquea* in Argentina is the Coligüe bamboo, which you probably know if you have visited the northern lake district in the Andean Patagonia.

When the bamboos die after flowering they leave a gap in the forest, but the seeds of fast germination form a new canebreak.

At least 18 species of birds are found associated with the bamboos in the type of jungle that Iguazú protects; amongst them some antbirds, a group which follows the Army ants (see page 70); or the Purple-winged Ground-Dove, which suffers lack-of food when the bamboos die after flowering. Other species of fauna associated to bamboos are the Cane Tree-frog, with suckers on their fingers to climb; a lizard that

climbs on the bamboos; and the Southern Bamboo Rat, with a long tail. Many of these species have night activity and you will most likely not be able to see them in a short visit to the National Park.

•••

Ferns

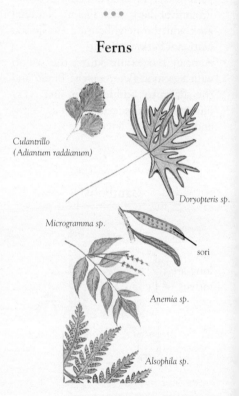

Culantrillo
(*Adiantum raddianum*)

Doryopteris sp.

Micrograma sp.

sori

Anemia sp.

Alsophila sp.

Several ferns, many of them growing on trees, can be seen in the paths of Iguazú. Leaves of some species of this type of plant are considered the largest and most complex in the Plant Kingdom, although there are also small ferns.

The main function of the leaves in a plant is generally to capture light for photosynthesis and to avoid evaporation. The simplest way to accomplish one thing or the other is in relation with the size of the leaf.

Ferns prefer dark and damp places, and when they grow on a tree, they do so in areas which are not exposed to the sun. This is why their leaves are very large compared to the rest of the plant: they offer a great surface to capture light and the risk of dehydrating is low in their environment. They can even afford to be thin and without protective layers to avoid evaporation. In this way the exchange of gases between the leaf and the environment is favoured.

If you have a fern at home, you surely know it must not receive intense light.

The big tree ferns up to 8 to 9 meters tall and with leaves of more than 2 m long are the most impressive representatives of Iguazú, although no sample can be seen from the Waterfalls Circuits.

These prehistoric plants develop in a comparable way to palm trees, with an only trunk from which the leaves protrude and which are replaced as they grow old. This way of growing might not favour the attachment of hanging plants or lianas or creepers on their trunk.

During the damp periods the stems and leaves of ferns develop very quickly. On the lower surface of the leaves a series of small circles called sori can be seen. These sori contain the spores which will be dispersed by the wind. They also reproduce from fragments of the rhizoid.

As they do not produce flowers, they are hardly related to animals: they do not need pollinators nor fruit dispersers. This is why they are not adapted to produce defensive toxins as the flowering plants do (for example, they do not produce alkaloids), although they produce some substances which are harmful for herbivores.

What can be seen is that they are hardly consumed: if you search for signs of attacks by herbivores, you will probably not find any.

• • •

Plants Growing on Other Plants (Epiphytes)

1) Bromeliads (Caraguatáes and Spanish Moss)

Caraguatá
(Aechmea sp.)

The Pineapple is probably the best known species of the family Bromelaceae, a group which includes 2,000 species found exclusively in America, of which almost half are epiphytes.

In the Iguazú jungle the epiphytic members of the family are the caraguatáes, with colourful inflorescences, and the Spanish Moss (*Tillandsia sp.*).

Leaves of the caraguatáes are long and narrow, in the form of a rosette, with grooves and sheathed in their base. This arrangement results, specially in the larger plants, in a tank which collects water and nutrients. This adaptation helps these epiphytes to grow without roots nourishing from the soil. An extraordinary case of "tank bromelia" was found in Panama, where a plant held 20 litres of stored water! In these large natural collectors, larvae and adults of several insects can live, and in the tropical forests there are even tadpoles and frogs (see page 77) that find refuge from many predators. Excretion of this fauna will dissolve in the water and is a source of nutrients. How do the liquids enter the plant? Mainly with the help of highly modified "hairs" (absorptive foliar trichomes) found in leaves, which also act as "unidirectional valves". The caraguatáes can solve their problems by capturing rain water, that is why they will not be found on trees in

dry environments. On the other hand, the claveles del aire are abundant in the jungle of Iguazú and also, for example, in the dry Chaco area.

Spanish moss
(*Tillandsia sp.*)

"What do they have that we don't?", the caraguatáes might ask themselves. To begin with, the Spanish moss are generally smaller and their leaves are covered by a dense mantle of absorbing hairs which allow them to capture humidity. But what is more important is that many species develop a method I shall call "night transport", which is possible thanks to adaptations of their metabolism.

What is "night transport"?

We know that carbon dioxide, which is captured through the stomata (pores for gaseous exchange), and light are essential for photosynthesis.
With "night transport" the Spanish moss can take up carbon dioxide during the night and close the stomata during the day to avoid loss of water by evaporation. Many epiphyte orchids also have this mechanism.

Butterflies and Hummingbirds are important pollinators of the Spanish Moss.

Coming back to the Bromelaceae, in the first part of the Macuco path and also in the Lower Circuit you will find samples belonging to a group which grows on the ground: the chaguares. They do not form a collecting tank as the caraguatáes do, but their shape allows rain water to concentrate at the base of the plant where it is absorbed by the roots. The borders of their leaves are serrated, which helps them to defend themselves from a greater number of land herbivores or avoids the damage caused tamping by animals, such as herds of peccaries.

The chaguares or Ihvirá are abundant in areas of Chaco, and many species develop "night transport".

2) Güembé (*Philodendron sp*)

Güembé

This beautiful plant belongs to the family Araceae. As many close relatives, they are used as ornamental plants in different countries. One of the interesting features of the Güembé is that small plants growing on the ground can send roots on the surface to a nearby tree and surround its trunk (different to the Guapoy who can start growing on a tree and send its roots to the ground). Some roots capture nutrients and humidity from the atmosphere and others are fixed to the host plant.

Both the Güembé and the Guapoy can also grow and live on the ground, without having to depend on another tree. The leaves of this plant are coriaceous and their size can be variable, according to the light and nutrients supply. The largest can be 80 cm long and present deep grooves.
If you look for the effect of herbivores on their leaves, most probably you will not find any. This is because the Güembé, as all the plants of the family Araceae, produce a defensive chemical substance (see page 42).

This plant flowers all year round, although its pollinators would only appear during spring. The small yellow berries mature in December and January and are food for many birds and some mammals, which would help in the seed dispersal.

Samples of great size, with hanging roots, can be seen clearly in the area of the Alvar Nuñez Cabeza de Vaca Fall. They can also be found in several parts of the Lower and Upper Circuits.
Some close relative used as an ornamental plant (for example several *Philodendron, Dieffenbachia, Anthurium*, etc.) could be growing at the entrance of your home, your study place or your work.

3) Orchids

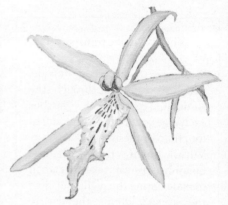

Orchid Miltonia flavescens

Orchids are probably the most diverse family of plants in the world. There are more than 20,000 species, with the greatest diversity in the tropical forests of America. Only in Colombia, some 3,000 species have been identified, a similar number to the orchids of all Africa, including Madagascar.

In Argentina, some 250 species have been identified, and after a long project done by the naturalist Andrés Johnson, about 85 species are known in the Iguazú National Park, 52 of which grow on other plants (epiphytes).

Apart from their abundance, the variation and specialisation of their sexual reproduction (see page 28) and their vegetative life are surprising. Surely, these facts are related. Remember: Evolution, change and diversity of species (see page 19).

As the epiphyte orchids must face the lack of water, many have bulb stems as an adaptation for storage and their leaves usually have a thicker cuticle than ground orchids. On the other hand, their roots can be covered by a layer of dead cells which retain humidity and gives them certain protection. They can also close their stomata (pores for gaseous exchange in plants) during the day to avoid evaporation and do the "night transport" already mentioned for the Spanish Moss (see page 58).

While the flowering of many species of orchids can last several weeks till they are pollinated, in some they only last one day!

Bees, flies, butterflies, moths, wasps and in a lesser degree birds, are their known pollinators.

Their flowers vary in shape and size, although they usually have three sepals and three petals. One of the petals is usually larger and colourful and is called labellum, and in many orchids it acts as a "landing site" for insects that arrive to feed on nectar.

In most cases the pollen of the flowers is concentrated in compact clusters called pollinias, that also have a type of adhesive gland to stick to the pollinator insects, ensuring that in the transport there is less loss of free pollen grains.

There are some species of orchids which can be self-fertilised (the pollen falls directly on the female structure of

Orchid with bulb stems

Wanted...

It is interesting to mention Darwin's article about self-fertilisation published in the English magazine *Gardeners and Agricultural Gazette*, June 1860.

After his famous trip around the world on the Beagle, the naturalist lived near Cambridge and for years observed a species of orchid which had self-fertilisation, without finding any pollinator insect.

The pollen was found in compact clusters (pollinias) with adhesive substances (which should have been to stick to insects).

Darwin was convinced that there should be crossing with flowers of other plants (the genes and Mendel's famous Laws were not known yet) and estimated that in another region of England the insects must act. The main concept of his article transmitted: "If anyone has seen an insect pollinating this orchid somewhere in England, please let me know!"

After pollination, the colours, the fragrances and all the means of seduction of the flowers disappear. They will produce a great number of very small seeds, which in most species are dispersed by wind.

In a fruit of an orchid of Central America 3.5 million seeds were found. As you can imagine, most seeds will not germinate, otherwise the world would be covered with orchids. To begin with, they must associate to mhycorrizae (fungi) which provide humidity and nutrients to their roots.

The droughts and frosts are limiting factors for the growth of these plants in Iguazú., that is why there are less orchids on the treetops (more exposed areas) than in the lower stratum.

In the riverine forests the climatic conditions are more constant, the trees are generally better exposed to light, to the currents of air and there is presence of nutrients and humidity for the growth of epiphytes.

About 50% of the orchids found in the Park are from the Waterfalls area, where the vegetation of the islands and river banks is influenced by the high relative humidity of the area.

In Iguazú, the flowering of different species occurs throughout the year. But many orchids are rare and not very frequent. Amongst the most common ones, we can find the *Oncidium* (Flor de Patito) which flowers from

the same flower, thus fertilising the ovule), without having to depend always on a pollinator . But the "price to pay" is very high: as the pollen grain and the ovule belong to the same plant it does not gain the genetic information from another individual. That is why pollinators are still needed to carry pollen to other plants.

December to March, and the *Miltonia flavescens* flowers between September and November.

It is important to consider that it is difficult to identify orchid species, specially if they are not flowering. But when you find one of these plants you will know some of the characteristics of the Family. It can be compared to the difference in observing a Picasso knowing a little bit of art and about the life of the painter, or to do so without having any idea about the subject.

On the other hand, their beauty and diversity has generated an important commerce of orchids in all the world. They are cultivated and species are crossed, producing new hybrids. But they are also taken from their natural environment, leaving many species in danger of becoming extinct.

In Venezuela they had the paradox that the orchid, the national symbol of the country, was in the verge of extinction because it blossomed only a few days before Mother's Day: it was exported in great volumes to the USA to be given as a gift of love.

Something similar occurs with the misunderstood love for animals which increases the dealing of pets such as tortoises, toucans and many other species which are taken from their natural environment. To appreciate the authentic significance of plants and animals and their relationships there is nothing better than observing them in their natural environment.

Fauna

Insects

Insects are the group of animals which includes the greatest amount of species and individuals in the world (see page 20).

Many take part in the pollination of plants, others help recycling by disintegrating organic matter, some regulate other species, some are a source of food for all types of organisms and many "co-operate" with other plants or animals in fascinating relationships.

In the jungle's network they are controlled by factors such as toxins produced by plants and by several predators. On the other hand, pests and plagues which affect man make their continuous control an important need.

Do not buy wild animals; neither buy plants, as orchids, that do not clearly show they have been cultivated.

63

In Iguazú they are more abundant during Spring, when they find a great supply of new leaves and flowers to feed on. Between May and June, instead, they would be scarce. Surely the absence of insects, more than any other animals, would produce a "halt in the activities" of the jungle.

Butterflies and Moths

The Iguazú Waterfalls Circuits (specially during Spring and Summer) show a large diversity of species and numbers of butterflies and moths concentrated in a small area.

They belong to the order Lepidoptera, a group which has approximately 150,000 known species, with a greater diversity in the tropical region of America. Less than 7% of the Lepidoptera are butterflies, which include the most colourful ones.

Their life cycle includes the stages of egg, caterpillar (larva), cocoon (pupa) and imago (adult).
In almost all cases the female lays her eggs on plants and fixes them with an adhesive substance.

Caterpillars eat so much that they are considered a mouth with mandibles in a head with a long, segmented body. Most of them are herbivores, that is why they need to process the defensive toxins produced by plants (see page 40).

In some groups of butterflies, caterpillars cut the vascular system of the leaf to stop the toxins transported through the sap from reaching the leaf. After a short period of time they will start feeding on that leaf.

Sphingidae caterpillar

Caterpillars also have several defence strategies against their predators. Stinging hairs, camouflages (some look like birds excretions on leaves, mosses, etc.) and toxic substances (they exhibit warning colours). The caterpillars of the Papilionidae butterflies, when bothered, show a couple of glands which liberate smelly chemicals in a defensive attitude.

During the pupa stage till the adult emerges, the butterfly suffers a number of changes. In some species of *Heliconius* butterflies the male mates with the female even before she emerges.
Some adult butterflies have only a few days of existence, while there are species of *Heliconius* which live up to

Heliconius butterfly

Not only with chemical compounds do the plants defend themselves from insects. In species of creepers of the genus *Passiflora* complex mechanism has been found to avoid the laying of eggs on them by the *Heliconius* butterflies. You will be surprised by the ins and outs of life. On one hand, the caterpillars of these butterflies are not affected by the toxins of the *Passiflora*. But these plants produce nectaries (type of glands) which offer nectar to certain wasps and ants which protect them from the caterpillars. On the other hand, some of these nectaries look like eggs of *Heliconius* (described as small, yellow spots). When a female *Heliconius* butterfly arrives to lay her eggs on this plant and sees the "false" eggs, she generally continues flying in search of another plant!

Other groups of plants have in their leaves' surface a layer of "small thorns" which might protect them from caterpillars, for example the stinging Giant Nettle (ortiga) of Iguazú. It has also been proved that the leaves with serrated edges hinder the movement of caterpillars as they feed.

three months, and the monarchs which can live two years.

Butterflies' wings are covered by scales and their colours can send warning messages to their predators or communication messages amongst them.

But they can also capture light to warm up their body. The Pieridae, for example, have a dark pigmentation in their wings which varies in size according to the season to absorb more or less light (the darker colours absorb more light). White wings instead, may act as a mirror, and the butterflies can direct more or less light to their body by changing the angle of their wings.

What do Lepidoptera feed on?
Most adult Lepidoptera have a sucking mouth which consists of a long coiled proboscis which can be uncoiled when they feed.

Many species suck the nectar from flowers, others the juice from rotten fruits, the sap from some plants or the fluids from dead animals. There are some moths that do not need to feed as they live on the food stored by the caterpillar.

Who feeds on Lepidoptera?

All the warning and camouflage adaptations (see page 47) that have arisen from the co-evolution give us a clear idea of the great pressure suffered by the butterflies and moths from their predators.

Amongst the invertebrates, their most common predators are spiders, wasps, ants, mantis, bumblebees and flies, and their main vertebrate predators are birds, but also mammals such as bats, lizards, etc.

What can we find in the Iguazú paths?

• Butterflies

Pieridae

Swallow-tails (Papilionidae)

"88"

The already mentioned *Morpho* and *Heliconius* (see page 44) are very stunning.
The Swallow-tails butterflies, many with attractive colours, belong to the family Papilionidae. It has been suggested that the prolongation of their wings would distract the possible predator if attacked, giving the butterflies a greater chance of escaping undamaged. The Pieridae have triangular wings and are yellow, orange or white. They are usually found in sunny places, they feed on nectar of flowers and most of them have a speedy flight.

The males of Pieridae and Papilionidae are usually found in mud or small puddles where they libate salts which allow them to maintain their ionic equilibrium. They would also obtain their raw material for the "chemical messengers" used to attract the females.

Butterflies of the family Nymphalidae are known by numbers, for example the "88", which has contrasting colours between the upper and underwings. The quick change of showing one or the other side of their wings would deceit their attacker. This effect has also been suggested in the genus *Morpho*. Another attractive Nymphalidae present in Iguazú is the *Doxocopa*.

• Moths

Sphingidae hawkmoth

During dawn and at night moths

appear. Although there are many moths that also have diurnal activity.

The light coloured and scented flowers captivate them in their search for nectar. Their great diversity make them play an important role in the pollination of flowers. Amongst them the Sphingidae feed while they are flying, as the hummingbirds do, using their long and thin extensible proboscis to reach the nectar. Many of them follow "fixed routes" to feed (see page 28).

The fast beating of the hummingbirds' wings, of nearly 80 beat frequencies per second, is fascinating. Well, there are some Sphingidae which have 200 beat frequencies per second and there are some bees, flies and other insects which reach up to 1,000 frequencies per second. Really these examples are not comparable. Amongst other things, the greater size of the hummingbird and the fact that it is "warm-blooded" determine that the laws of physics limit in a different way. And also, for our perception of reality, there is nothing like a hummingbird! (although, I'm sure entomologists think different).

The Sphingidae are well represented in the Iguazú National Park. They can be distinguished by their large body, rounded head and a conical abdomen ending in a tip. The front wings have a triangular shape and are longer than wide, while the back wings are small and their colours are not very attractive.

Another fantastic feature of moths is the males sensitivity to detect the chemical messengers of sexual attraction (pheromones), liberated to the air by females.

Numbers are sometimes good enough just to say amazing!

Pay attention to the following: If a female silk butterfly liberates 100,000 times less than 1 gram of pheromones (chemical messenger) per hour, it can attract a male at a distance of nearly two kilometres.

And how does the male fall to the seduction of the "perfume"?

It has two antennae of an amazing sensitivity. In each one of them there are 1,700 sensitive hairs and each hair has 2,600 olfactory pores.

The family Saturnidae (an example is the silk moth of the Old World) has stunning representatives in Iguazú.

During the day, moths try to pass unnoticed on the trunks, branches or leaves of trees. There are too many hunters in search of nutritive food. But at night (specially during spring and summer) you only need to get close to a strong lamp to find them. That is a good opportunity to have a magnifying glass at hand to observe the "hairy" antennae (a characteristic which distinguishes them from the antennae of butterflies) in the moths on the ground, or the length of their proboscis which allows them to reach the nectar of flowers (although in some it is atrophied). You will also find a diversity of beetles and other insects.

Now then, why are moths attracted to light?
It has been suggested that artificial light confuses them, as they might find their orientation thanks to bright points (such as stars or the Moon), maintaining their angle of flight according to them. This means they would have the capacity of remembering and learning, as they move in an oriented way they would do it in relationship to something. For example, in relation to the plants where they lay their eggs or where they feed; a step further than the simple attraction to a flower by chemical compounds.

Correct Prediction

The naturalist Charles Darwin received information from explorations in Madagascar about a flower whose corolla formed a tube longer than 30 cm. Some animal was bound to pollinate it, but as no species was found, he predicted that there must be a moth with a proboscis at least as long as the tube of the flower. As a matter of fact, it was discovered forty years later! It belongs to the family Sphingidae.

Invincible Armies: ants

Be impressed by these examples:
• In a 1987 report by Edward O. Wilson (a biologist with a distinguished research career dealing mostly with social insects) there is surprising data. In only one tree in the Tambopata reserve (area of the Peruvian Amazon which protects one of the jungles with the greatest diversity) he identified 43 species of ants, a similar number to all the species found in Great Britain.
Less than five years later, in the Manu National Park in Peru, 72 species of ants were identified in only one tree! (remember the concept of species; see page 18).
• After only one copulation, the queen Fire ant of the genus *Solenopsis* (which is found in South America and is present in Argentina with different species) uses sperms in a regulated way to produce between 60 and 200 eggs per hour. In one day she lays the equivalent to her weight, which she recovers by the continuous feeding she receives from the worker ants.
She can keep this rhythm of egg laying for six years, which would mean 3,000,000 eggs in a lifetime.

• • •

Leaf-cutter ants

Leaf-cutter ant

You might have encountered these hard working ants in your garden, and there must be many of you who, in your childhood, felt curious about them.

Leaf-cutters are represented by 200 species distributed from the south of the USA to Northern Patagonia. Of them, 37 species specialise in leaves, while others transport seeds and different organic matter.

They form societies of millions of individuals and their importance as herbivores in the natural environment is greater than any other animal.

In Barro Colorado Island, in Panama, (see page 15) it has been calculated that Leaf-cutter Ants consume 300 Kg of vegetation per hectare per year, an amount greater than the total consumption of all the vertebrate herbivores of that environment. In tropical forests of America their consumption is 15% of the total production of leaves.

All the plant material they transport is used to feed certain fungi they grow in their nests and which represent their source of food (they also ingest sap from plants).

The relationship between fungi and ants results from their co-evolution (see page 19) and their dependency makes them inseparable. That is why the queen which starts a colony must take a fungus sample for its growth. DNA research concludes that these ants have been depending on the same strain of fungus for about 30 millions years.

The worker ants in charge of finding leaves are guided in their path by their sense of smell and touch rather than by their sight. They must choose the plants they shall visit, as some have terpenoids, substances which are toxic for the fungus (see page 41).

With "marked" individuals it was proved that they keep their path to a certain plant and do not accept any alternative path which might be offered.

In some species, apart from the transporter worker ants, there are other smaller workers that travel on the pieces of leaves. Their goal is to avoid the attack of certain flies which try to lay their eggs in the hard working worker ant. When they arrive to the colony they find the soldier ants guarding the entrance, these are larger female ants whose function is to defend.

Once inside, they hand their load to other workers which are smaller, who chew and triturate the pieces of leaves and mix them with their excrement and saliva. The excrement supplies the enzymes needed by the fungus to be able to digest certain proteins.

The fungi would be a great filter of plant toxins: the ants are the group of studied insects which predates the greatest diversity of plants!

Wherever you find an army of Leaf-cutter ants moving in a line, it is worth while to stop to observe them. Try to detect the differences in size between the ants, between the size or type of material they transport, and the arrival of an occasional insect which eats them.

"To do this I need to lay on the ground, or at least bend down", you might think. Of course, and if anybody passes by, you might tell them about these tiny creatures which together can "move mountains".

Tiger ants

Tiger ant

If in your walks in Iguazú you find some black ants of a great size, which can be up to 2.5 cm. long, do not try to catch them.

These are the Tiger ants, carnivores with potent mandibles and with a sting which causes great swelling and intense pain in man. Even the Collared Anteater, who bases an important part of its diet on ants and termites, avoids eating them.

They form small colonies and the different castes are not very noticeable. They are more primitive than the Leaf-cutter and Army ants. They move on their own and their touch and olfactory senses guide them in their search for food, which includes several invertebrates.

Army ants

Army ant

Occasionally, great hoards of voracious carnivorous ants appear in the different forests in America, including Iguazú. These are the Army ants represented in our continent by some 150 species.

Their columns can have several dozen metres long and, as they advance, they can even climb trees. Other ants, wasps, termites, several insects and spiders are their victims. There are some wasps and beetles which have created an "immunity countersign"(smells similar to those of the Army ants) and they are not attacked.

Amongst the vertebrates, only small individuals which can not escape or which are injured, would be the victims of these ants (frogs, lizards and possibly the fledglings of some birds).

The Army ants are guided by their sense of touch and smell, as their sight can not form images and they only respond to the stimuli of light.

There are workers of different sizes specialised in tearing to pieces their prey, while the soldier ants, larger in size, act in the defence and they have a characteristic long mandible which resembles a hook.

Different birds usually follow these armies to eat several insects which escape from the Army ants. This occurs in Iguazú with the antbirds and some woodcreepers (see page 84), amongst others.

There are butterflies which also follow the Army ants, but their interest is in the excrement of the birds that follow these ants, which is rich in nitrogen. They belong to the family Ithomiidae, toxic butterflies (with warning colours; see page 44) which do not result in an interesting bite.

The sudden advances of these "nomad" ants respond to an internal clock which depends partially on the laying of eggs of the queen. When it is time to feed the larvae, this will start a new raid.

As they advance through the Misiones jungle they can invade ranches and homes. You might find a column of Army Ants in your visit to Iguazú.

• • •

Beetles: Rhinoceros Beetle and Dung Beetle

Beetles belong to the order Coleoptera, the most diversified amongst the insects. With more than 350,000 known species (all birds and mammals together do not add up to 14,000 species), their diversity and abundance motivated the naturalist and philosopher J. B. S. Haldane to comment that God must have had "an inordinate fondness for beetles"...

They have a first pair of chitinized wings, the elytra. Underneath, the second pair of wings is protected. This pair is membranous and allows many species to fly.

Rhinoceros beetle

Rhinoceros beetles have a great number of species and are found in many habitats of the world. They are called like this after the showy horn of the males. It is used for the defence of feeding areas or in battles between males to obtain the favours of the females. The adults have masticating mouthparts adapted to a herbivorous diet.

Tropical species of a greater size of this type of beetle have larvae which need trunks of great diameter to grow. But the intensive and selective cut-down has left few large trees, affecting the reproductive cycle of these insects, amongst many other problems.

Let's see what happens in the jungle to the dead vertebrates. The carcasses of Brown Capuchin Monkeys or of brockets for example, can be food for vultures (see page 96) and a diversity of insects, such as flies and carnivorous ants. In a short period of time they shall leave only bones, and the organic matter will enter the cycle of life once more.

Excrement of the vertebrates is recycled in a short period of time and is nutritive food for several organisms.

The dung of the Tapir or of peccaries can be a deposit for eggs of flies and attraction for several beetles, many of bright colours.

Some of these Dung Beetles make a ball of excrement which can be larger than their own size and they later push it as best as they can: with their hind legs. When they are in a couple, the male pushes and the female rotates on the ball in order not to fall. Later they shall lay their eggs and will bury the excrement, creating the conditions for

Dung beetles

the development of larvae, and helping in an indirect form to the fertilisation of the soil.

Rhinoceros and Dung Beetles can be observed in the paths of Iguazú, specially during Spring and Summer. They represent a very small sample of the practically unknown world of the insects in the jungle.

Harlequin Beetle

Harlequin beetle

The Harlequin beetle is a beautiful Coleoptera of the family Cerambicidae, group which includes the wood drillers. This species ranges from Mexico to Argentina, and is present in Iguazú. Their body can be 7 cm long although with their long hind legs extended they can reach 20 cm. Their attractive colours combine black, yellow and red. They have a life cycle representative of many insects whose larvae grow inside trees, which they can greatly harm or even kill. The woodpeckers and hummingbirds eat the larvae and help in the control of these insects.

The female Harlequin Beetle lays her

eggs on trees and makes a hole in the bark. This will be the means of entering the tree for the larvae, which will develop in a few days.

In Costa Rica it was determined that during the following seven months the larvae drill a series of galleries with holes to the exterior and after spend about four months in the stage of pupa(cocoon). Closing the cycle, the adult emerges from the inside of the tree and feeds on the sap of several plants. It is more probable to find the adult Harlequin Beetle during the night with the help of a source of light. Drilled galleries by this and other members of the family Cerambicidae damage the plants and usually cause the breaking of branches or small trunks of several centimetres in diameter.

If in your walks you observe carefully, you will be able to find the cut stems of plants (of several cm wide) with their top ending in a tip as if a sharpener or knife had been used.

•••

Dragonflies and their Relatives

Dragonfly

73

"The number of dragonflies in the sky is a signal of rain!". This is a phrase very much heard of precisely when there are many dragonflies flying around. But if you remember what I am going to tell you next, whenever you hear this phrase you will be able to add: "yes, it is true and..."

These insects belong to the order Odonata, a group which originated nearly 300 million years ago. Amongst their extinguished relatives there were species with a wing-span larger than that of the Rufous Ovenbird, the national bird of Argentina.

They are efficient hunters, with a very sharp eye-sight which has a field of vision of 360° and wings moved by muscles which can represent almost half the body mass.

Although their wings beat between 30 and 50 times per second, they are amongst the fastest insects in the world, with speeds of up to nearly 80 km/hr.

If the adults represent the fighters of the air, the aquatic larvae are the "sharks" of the water. They attack from small fish and tadpoles (see page 77) to larvae of different insects. They live in the water from one to three years before becoming adults.

Which are the weapons of these larvae? Many use the anal jet propulsion send-ing out water, being able to attack or escape quickly and causing surprise. On the other hand, they have a hanging lip with two hooks at each end, which unfolds quickly and allows it to catch its prey.

In Iguazú there are very colourful adults, and if for example you look in the swimming pool of a hotel you will probably find some trapped in the water. With a magnifying glass you will be able to see their large compound eyes, the complex supporting web of the two pairs of fragile wings and the mandibles with which it traps mosquitoes, flies and other insects.

• • •

Camoati Paper Wasps

Camoati Paper Wasp and its hive

The Camoati Paper wasps are a group of social insects of the genus *Polybia*, with a wide distribution in Central and South America.

They build big nests of plant fibres on the branches of trees or shrubs, generally close to gaps in the jungle. The size and shape of the nests can be variable, and they usually have little cones on the outer surface. Cellulose is the main component of plant fibres and to advertise it one might say: "This substance is light, easy to transport, resistant and does not dissolve in rain water (although this is not always so); you will never be able to find anything like it!".

Many animals "know this", as there are termites, several bees, and other wasps as the Lechiguana who also build their nests of plant fibres.

Inside the nest, the female Camoati Paper wasps make cells in different levels and fill them with eggs. They feed on the larvae of butterflies and beetles and feed the developing generation with termites.
As time goes by, aggressiveness and cannibalism arises in the society and finally some young queens and workers abandon the nest to form a new colony.

Camoati Paper wasps are black and yellow, and hardly reach one centimetre long. Their nests are attacked by mammals such as the Collared Anteater and some birds; the wasps are ready to defend themselves with their sting. Walking on the Iguazú paths you might be able to see their showy nests.

• • •

Melipona Stingless Bees

Melipona in front of the nest entrance

The Melipona Stingless bees form part of the family Apidae, the honey bees (originally from Europe), known by everybody and the euglossine bees (see page 28) also belong to this family.
Melipona bees are small and do not have a sting. They make their nests in the base of hollow trunks, safe from predators.

Their nests are made of wax or resin and have special containers for honey, pollen and little cells for the offspring. They communicate with the outside through a little yellow vertical tube, which can be detected in Iguazú with careful observation. Close to the entrance you might see some individuals flying or posing on the tube.

— • —

Spiders

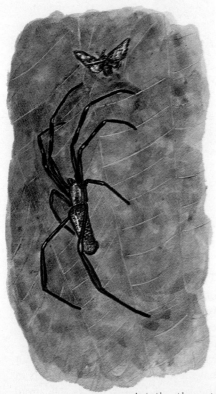

Argiopidae with a prey in its web

The spiders are a very diverse group and are present in almost all the land habitats of the world. What probably calls more our attention is their silk production, material which hardens in contact with air.

Some male spiders use them to discharge their sperms and to later submerge their "arms" in order to fertilise the ova of the females which in many cases are also deposited on the webs. They are carnivores and play an important role in the insects' control. Soon after they are born, many spiders disperse, lowering the competition amongst them. Juveniles of several groups climb to the top of plants and extend a thin silk thread which acts as a "sail" in order to be transported by the wind. Of course there are always some risks, as being trapped in the air by the Great Dusky Swifts (see page 83) of Iguazú.

The spinning of their web is what most calls our attention, although not all spiders do it and on the other hand some webs are nearly invisible so researchers need to use starch to "develop" them. Some are irregular and others have very elaborated designs. In the Waterfalls Circuits different examples can be found. Silk threads used to trap a prey are sticky, but the external threads used as support are not sticky. Their resistance and elasticity are amazing. A thread with a diameter of only 0.1 mm can hold a weight of 80 grams and can be stretched up to a 20 % of its length before it breaks.

Spiders lie in wait and detect the slightest vibrations in their web, which can be caused by their couple or a prey, generally insects.

Although there are mosquitoes whose legs are so thin that they are not detected by the spiders. Some butterflies can escape as they lose the scales of their wings, which stay stuck to the web. Other butterflies are so distasteful that the spiders let them free.

But when the victim is "tasty", the spiders usually inject their poison to immobilise it and enzymes to digest its tissues, as spiders do not feed on solid food. If the prey is of a considerable size, they can wrap it up in more silk. Spiders ingest the fluids, sometimes only leaving the exoskeleton ("carapace") of the insects, as you might see in your walks.

There are others who benefit from the spider webs. For example, the hummingbirds use the silk to stick together the materials of their nests. But this might be a dangerous move. In Brazil, it has been observed how a Black Jacobin (hummingbirds not very common in Iguazú) became a mortal victim of a spider when trying to obtain silk.

On the other hand, spiders of long, black legs which are found in Iguazú,

are hunted and paralysed by a large wasp called San Jorge, which offers them to its larvae as food.

The rhythms of life and death are Laws of Nature.

— • —

Amphibians and Reptiles

A Jumping Life:
Frogs and Toads

Why do most frogs and toads jump instead of running?
To begin with, the length and power of the muscles of their hind legs is amazing, specially in frogs. These are fundamental for swimming quickly and for jumping.
It has been suggested that this means of locomotion on land is an evolutionary answer to the need of escaping in a vegetation environment where the movement is complicated.

Nosed Climber Tree-frog

77

Another suggested advantage is that the jump leaves irregular olfactory traces, which probably helps to confuse predators as snakes. It has also been convenient in order to move between trees, habitat of many species of frogs.

On the other hand, they are very vulnerable during their aquatic life (where most species lay their eggs and develop into tadpoles), specially in the jungle with its countless predators.

In the jungle, the evolutionary tendency has led to the independence of water environments. An amazing example is that of a species of tropical frog of the genus *Dendrobates* whose females transport their tadpoles on their backs and climb trees. They deposit their offspring in different tanks of bromeliads (see page 57) to later descend to the ground.
Every now and then they climb up to control if their tadpoles are still alive and if this is the case, they feed them by laying sterile eggs!
In Iguazú there are no species of frogs so specialised. But it is effective to lay eggs in temporary pools, avoiding the predators as fishes found in permanent waters. Although the larvae of dragonflies (see page 73) are voracious hunters of tadpoles, the herons and other birds (see page 97) also find in them a good bite. Many frogs lay their eggs on a mass of foam, others on vegetation and their tadpoles will be aquatic.

In the San Martin Island, the spray of the San Martin Fall can fill a series of pools found just before reaching the balcony of the fall. Tadpoles will develop there in Spring and in Summer and will feed on algae and plants.

The Tree frog *Hyla faber* is a more specialised case found in Iguazú: it constructs a dam to keep the liquid in a pool where it lays its eggs. The frogs of the family Hylidae can live on trees or on the ground and generally have adhesive disks on the tips of their fingers which allow them to climb easily. In Iguazú, several Hylidae frogs find refuge in leaves of the Bromelaceae, where they compact to camouflage.

In several species of amphibians, chemical defences are secreted by glands on the skin. There are extreme cases, as the *Dendrobates*, tropical forests frogs, whose skin has a potent toxin. It was used by the Chocó Indians in Colombia to make their arrows into mortal weapons.

What colour are the *Dendrobates*?
Very colourful, as the ones that combine black and red, to give their warn-

Toad catching an insect

ing message (see page 44). It has been suggested that the toxicity of these frogs evolutionary developed in relation to their need of remaining still, thus exposed to their predators, while they lie in ambush for their prey, ants and termites.

The Rococo Toad is a species common in Iguazú. It has glands that produce a milky substance which is toxic if absorbed by an animal's mucus, but does not penetrate skin (for example ours). This toad can also swell to avoid being eaten by some snakes.

After it rains and at dusk on warm days it is possible to see toads capturing insects with their fast and long tongue. Because of their diet, they are important insect controllers.

The Nosed Climber Tree-frog is amongst the most common frogs in Iguazú, and it can even be found in bathrooms of hotels or in water tanks of homes.

Frogs and toads are basically animals with dawn and night activity, thus not easy to detect. Never-the-less, in Spring and Summer the evening croaking of the males calling the females is a sign of their presence and contributes to the magic and charm of the jungle.

In Iguazú some pools of considerable surface can hold eight species of frogs, each occupying a different niche and developing in Spring their own nuptial song. These "love calls" have their risk:

they tell predators, as bats, their position.

• • •

Tropidurus lizards

Tropidurus lizard

The climbing lizards of the genus *Tropidurus* are abundant in Iguazú and can be found in most paths of the Waterfall Circuits. Their presence is related to the areas of exposed rocks which are abundant in the Circuits.

As they are "cold-blooded" and small animals they can hunt with fast but short-lasting movements. They feed on insects as butterflies, on spiders, worms, other invertebrates and in a lesser amount on plants. They have been seen entering the nests of the Great Dusky Swift, probably to predate their eggs.

What type of defence do they have against their predators?
On one hand they camouflage very

well, and you will notice this when you see them on the rocks or trunks they climb on to.

As many other saurians, they can lose their tail, a defensive adaptation which allows them a greater possibility of escaping. The separation is produced in areas of the skeleton which are less resistant and the loose tail can violently shake due to uncontrolled nervous impulses. This can result in a distraction for the predator.

Although the tail grows again, it does not reach the original size. Losing the tail has its disadvantages, as it would act in the balance when climbing trees and in its base fats can be deposited.

• • •

Tegu Lizard

Tegu lizard

The Tegu Lizard is the largest lizard in Argentina, with reported individuals up to 1.5 m long, although they are generally less than one meter long.

Does their large size allow them to feed on small and fast prey as most lizards do? It would be very inefficient due to the excessive waste of energy in its movements. They have a varied diet which changes according to their age. The juveniles feed on snails, spiders, wasps and a diversity of invertebrates. They later add fruits and with blows of the tail they open the nests of lechiguana wasps to obtain their honey.

As it grows, it varies its diet and its teeth. Its doublechin becomes very noticeable due to the development of the chewing muscles, specially in males. Their food will consist of small vertebrates such as frogs, fish, rodents, eggs and even carcasses.

Is the amount of food ingested by the Tegu Lizard similar to that ingested by a cat of its same weight? The answer is no. The main difference is that the lizard (as all reptiles and amphibians) is a "cold-blooded" animal, different to the cat which is a "warm-blooded" animal. The latter uses 90% of the food energy to maintain its body temperature. Instead the Tegu lizard can live a long time with only one large prey. In low temperatures it remains inactive.

Because of its large size, it has less predators than the lizards, thus can spend more time warming up in the sun.

The mating season is in October and

November. The fertilised female uses some natural cave or digs one of her own to deposit her eggs, which can be more than 30.

Snakes and birds of prey are predators of the small juveniles, of approximately 20 cm long, while the Jaguar and the Ocelot are some of the predators of adults.

As a mechanism of defence, they use their teeth and they can also lose their tail, which does not reach its normal size if it grows again.

The Tegu Lizard can be found during Spring and Summer in the area of the Dos Hermanas Fall or on the road leading to the footbridge of the Devil's Throat (Garganta del Diablo), amongst other places. During Winter they take refuge in caves to hibernate and even on warm days it is improbable to see them.

• • •

Broad-snouted Caiman

The rare Broad-snouted Caiman has a nesting population in the islands of the Upper Iguazú River. It is probably the most important in Misiones and is one of the few populations reported for a jungle environment in Argentina.

It feeds of snails, fish, frogs, toads, snakes and in some occasions it captures small mammals such as rodents. With its large mouth with two layers of sharp teeth, it holds its prey to later swallow it with the help of its tongue. It generally hunts during the night and can be seen in the dark by flashing a torch in its eyes, which shine with a red light. It mates in Spring and the female builds her nest with plant materials. After laying her eggs, which can be between 20 and 40, the female covers them with debris, whose decomposition will generate heat for the incubation. The eggs hatch after one or two months and the vulnerable offspring, about 25 cm long, will remain with their mother for about a year. They will reach their sexual maturity in 5 or 6 years.

Several birds, as the storks, herons and birds of prey, and some mammals as otters and the Crab-eating Fox are predators of eggs and newly born. It is known that the adult Broad-snouted

Broad-snouted Caiman

Caiman can be attacked by Jaguars (see page 106).

The chances of finding them depend on the season of the year. During the cold days of Winter they are usually submerged. On the sunny days of Spring and Autumn it is more probable that they come out to warm up and during the hot days of Summer, they will most probably remain in the water.

They can be seen in the small dam of the old quarry found near the access to the Park.

According to fossil records, this species would have originated two million years ago. A few decades of intensive hunting in search of its hide placed it in the border of extinction. Luckily, the measures taken for its conservation and a fall in the price of its skin gave populations a possibility to recover.

— • —

Birds

There are 448 species of birds registered in the Iguazú National Park (nearly 50% of the birds identified in Argentina), although the presence of 20 species needed to be confirmed. From the Inventory of the Birds of Iguazú National Park (1996), we know that 305 live in the Park all year round; 50 are species which visit the Park in Spring and Summer; 6 visit it in Winter and 58 are occasional visitors (the remaining species are mentioned as of unknown seasonal presence).

The birds are key parts in the dynamic jig-saw puzzle of the jungle, with groups which act in the pollination of flowers (the hummingbirds), the dispersion of fruits, the "cleaning" of carcasses (vultures) and the predator-prey game. To know the conservation situation of a jungle or another natural environment one of the methods mostly used is to survey the birds, specially certain species.

Birdwatching requires the use of binoculars, and for their identification the help of a field guide (*Birds of Argentina & Uruguay* by T. Narosky & D. Yzurieta; *Birds of Southern South America and Antarctica* by Martín R. de la Peña & Maurice Rumboll)
I will mention only some of the groups, mainly those of greater possibility of their observation in the Waterfalls Circuits.
They are classified here according to their main (but not only) source of food.

Birds that Feed on Insects:

Swifts

Which are the birds which can be observed in a greater number and with a greater frequency all year round in a tour through the Iguazú Waterfalls Circuits?

Without any doubt: the Swifts, who at first sight resemble swallows. They have a small and aerodynamic body, very short legs and long, narrow and pointed wings. These indefatigable fliers hunt insects and even small spiders in the air (see page 76) with their big mouth and small beak. While they fly they can mate or take materials for their nests, such as feathers or pieces of moss in suspension. They mate in the hollows of basalt where they make their nests, as is the case of the Great Dusky Swift.

Great Dusky Swift

They do not perch on branches or on the ground, but they can be seen on the rocky vertical walls, which they catch on to by their nails and they find support with the callosities of their tarsals and their rigid tail. They build their nests by sticking them with saliva on cavities or projections of rocky walls.

The most common species in Iguazú is the Great Dusky Swift, with greyish-brown plumage. The National Park is one of the few nesting places of this species in Misiones (it was chosen as an emblem for the Park).

It is possible to observe the Great Dusky Swift at rest from the paths of the Lower Circuit. You have to look between the rocks of the Alvar Nuñez

Great Dusky Swift

Cabeza de Vaca Fall, in the Dos Hermanas Fall or in the walls of the Bossetti Fall.

It has been proved that on rainy days they remain in their resting places, while on sunny days they start their activity very early. In the area of the Devil's Throat they can form flocks of up to 3,000 individuals.

In the Falls of greater volume of water, it is surprising to see how they appear and disappear behind the curtain of water in their fast flight while passing through clearings or gaps.

They usually build their nests behind the waterfall, safe from their predators. Although not all is a "rose in the garden of Eden", sometimes "thorns" are felt. In Brazil there are reports of swifts which have been trapped by the waterfalls, suffering injuries or even death.

It has also been suggested that the laying of few eggs and the long incubating period are a consequence of the low temperatures and high humidity their nests are exposed to. An adaptation to this is the protective plumage the fledglings of some swifts acquire.

Woodpeckers and Woodcreepers

Woodpeckers are an arboreal group of birds and are very diverse and present in many different environments of the world. In Argentina, the southernmost species reaches down to Tierra del Fuego.

They use their thick bill to make holes in wood and to extract with their sticky tongue a diversity of insects, specially larvae. In Iguazú they help to regulate the insect population, as they ingest, for example, the larvae of Harlequin beetle and its relatives (see page 72), or several ants.

It has been suggested that their type of food is an evolutionary specialisation which ensures them abundant food very few animals have access to, although it could mean a great cost of energy to obtain it.

Their legs (with two fingers facing for-

Robust Woodpecker

ward and two facing back), and the support of their rigid tail, give them good support on trunks. There are few species that feed on the ground, and many complement their diet with seeds and fruits.

They also make holes in trees to build their nests. Many start two or three cavities to finally choose one of them. They leave in this way "half built" nests to cover the possibility of an urgent move. For example, in the case of being displaced by birds who search for hollows, as can be the case of some toucans, parrots or tytiras.
The holes which have already been used as woodpeckers' nests are not usually chosen by other birds as they have parasite insects and worms which might attack the fledglings of the new tenant.

They are quite territorial and it is possible to find them in couples or alone. The drumming they produce with their bill on the trunk also allows them to establish their territory, maintain communication with their couple and maybe to find a good trunk in which to build their nests. The drumming gives us a good indication of their presence, and for an expert ear, the possibility of identifying the species.

In Iguazú, the most showy are the woodpeckers of red hoods and a body with mainly black plumage, as the Robust Woodpecker. Piculets, as the Ochre-collared Piculet, are small and hardly climb. They generally move through small branches and produce a high-pitched noise.

On the other hand, woodcreepers are a group of birds which resemble woodpeckers as they move on the tree trunks, but they are related to the Ovenbird's family (Furnariidae). They have three fingers facing forward and one facing back and their tail is not so rigid as the woodpecker's. They usually move in spirals around the trunk and with their bill they take food out of grooves, hanging plants or fissures in the bark.
Their plumage is usually dark brown or reddish. The noticeable variation in shape and size of their bill would allow them feeding specialisation of different insects and thus decrease the competition amongst species. For example, the Black-billed Scythebill, not common in Iguazú, has a thin bill up to 6.5 cm long.
These birds also make their nests in hollows of trees and there have been cases of aggressiveness toward woodpeckers to steal their nests.

Tyrants

The Tyrants (family Tyrannidae), are a group exclusive of America formed by more than 380 species of birds (about 130 in Argentina), which base their diet on insects. Such a diversity could have arisen in relation to the evolu-

Boat-billed Flycatcher

Great Kiskadee

Long-tailed Tyrant

tionary specialisation to capture their main prey. Since a long time ago, they have had to deal with insects of several shapes and sizes, discover many camouflages and hiding places (see page 47), follow fast fliers (see page 73) and learn the warning messages of the toxic insects, as the colours of many butterflies (see page 44).

These hunting birds have a very good eyesight and must be very fast to obtain their food. To be more efficient, their brain captures "signals" of shapes, colours, sizes and other variables of what might be their possible preys.
They have developed different hunting mechanisms and the variation in the size of their bodies and bills (generally ending in a hook) decreases the competition among species.
For example, the Great Kiskadee, the Boat-billed Flycatcher, the Vermilion-crowned Flycatcher and the Three-striped Flycatcher (all species which make nests in Iguazú and whose plumage is similar), have great differences in the size of their body and in the shape and size of their bill. The Boat-billed Flycatcher has almost half the size of the Great Kiskadee, but its bill is quite a lot wider. You will have to observe with detail in order to know which one you are seeing (with binoculars, of course). To identify their singing will help you to differentiate them more easily.

The Fork-tailed Flycatcher, although not a typical bird of the jungle, can be seen during Spring and Summer in the Waterfalls area. The opening and closing of its long tail (up to 30 cm long in males) gives it great manoeuvrability to hunt in the air.

The male of the Long-tailed Tyrant has two central tailfeathers up to 9 cm long and has a distinctive black plumage and its crown and neck are white.

The Tropical Kingbird is common in Iguazú in Spring and in Summer. They perch in visible places and from there they perform spectacular insect air hunting, although they also feed on fruits. Their long and pointed wings, their wide and triangular tail and their short legs facilitate the manoeuvrability in the air.

The adult of the Piratic Flycatcher is a member of this group that has specialised in eating fruit. It steals the nests of the Red-rumped Cacique (see page 95) and other species of birds.

Tytiras are birds which were related to the Cotingas till not so long ago. They usually perch on high and visible branches. Their plumage is white and their head, tail and part of their wings are black, which makes them easy to be identified. Fruits are the main food supply.

• • •

Birds that Feed on Nectar

Hummingbirds

Black-breasted Plovercrest

"To write a book, plant a tree and have a child", this is said to be what one has do to in life to feel fulfilled. It would not be out of place to add "observe a hovering hummingbird, libating a flower" (see page 27).

It might sound exaggerated, but to observe these birds is a "gift" of Nature which can not be missed. Even better if the observation can be done with binoculars, but this could not be mentioned in the phrase as it would not have sounded well.

The hummingbirds are a group formed by approximately 320 species exclusive of America. In the Iguazú National Park there are 13 recorded species, although some have only been observed in a few occasions.

They develop their fantastic flight with

a high beat frequency of their wings. From 14 frequencies per second in the slowest ones, about 37 in the Glittering-bellied Emerald, up to 80 frequencies per second in the Amethyst Woodstar (species hardly ever seen in Iguazú).

The sequence of movements of the wings with such a fast beating has been compared to that of the Hawkmoths, which also feed in flight (see page 66).

The energy they use is recovered thanks to the caloric nectar of the flowers and they also feed on insects to obtain proteins. In individuals of some species, it has been recorded that they ingest up to 6 times their weight in a day, while the Glittering-bellied Emerald ingests more than twice its weight on an active day.

What feeding strategies do hummingbirds have?
Flowers of at least 30 families of plants are visited by hummingbirds, who turn out to be good pollinators of many of them, such as the Bromelaceae (see page 26).

Most species defend aggressively their feeding area, which includes flowers of different plants. They are "generalists" and can have a lot of competition.

Some species can introduce their head and even part of their body in a flower, but later, with the many baths they take a day, they clean their plumage.

The length of their bill and probably the sustenance capacity of the wings limit which flowers they can access for nectar. The bill of the Sword-bill Hummingbird (a species of the semihumid to wet mountain forests of Venezuela, Ecuador, Peru and Bolivia) can reach 11 cm long, to which we must add the length of the tongue! It feeds with a system of fixed routes (see page 28) and is not very aggressive. Which other animal will be able to reach the nectar found in the base of a corolla which is as long as its bill?
Maybe only some moths, which can have a very long proboscis (see page 66), or nectivorous bats.

The small size and intense movement of the hummingbirds brings up an interesting question: Could they live feeding only on leaves if they had a bill and digestive system adapted to do so? Let's see the following comparative chart of relative caloric values of different foods, considering the relative value of leaves as 1.

Relative Caloric Values:	
leaves	1.0
fruit	2.0
insects	3.7
nectar	7.2
pollen	7.6
seeds	16.0

(Source: relative conversion from the maximum real values taken from Margalef R., 1986).

It is clear that the leaves are not a good source of energy. Specially if we consider the fact that in most cases cellulose can not be digested.

Conclusion: The day would not be long enough for the hummingbirds to eat the amount of leaves they would need in order to maintain their metabolism.

The truth is, that in very few "warm-blooded" animals (this is not the proper term to be used, but it helps to describe mammals and birds) can survive only on leaves. Those that do so, have a very low metabolism and can not be small. An example of the Central and South American forests are the sloths (they are not present in Argentina), the slowest mammals on Earth: they move at an average of less than three meters per minute!

• • •

Birds that Feed on Fruits and Seeds

Several birds base their diet on fruits, a food which is available nearly all year round. There are Strangler figs (see page 52) trees with fruit in the four seasons, and others as the Pindó palm, the Palmito and Aguay offer fruit during the season of greater scarcity: Winter. Although if we consider a given time, the abundance of fruit is not very high. Remember that in the jungle the rule is: "there are many species but a few samples of each".

The movement in groups of the frugivorous birds, specially during winter, would give them a better chance of finding food. Once the food is detected a racket starts, but there is usually enough for all.
Parrots act more like predators of seeds, but the other frugivores mentioned are good dispersers of seeds (see page 34). Many complement their diet with several invertebrates.

Parrots

The parrots are gregarious and noisy birds, specially abundant in the Southern Hemisphere.

They are good fliers and have adaptations to live on trees. Their legs are short, with two fingers facing forward and two facing back, and their powerful beak acts as a hook. In most species the dominant green colour they present with their folded wings helps them to be confused with the foliage.

They are very sensitive to taste, and it is astonishing how they can handle food. They usually use their legs to hold the fruit while with their strong tongue they take out the pulp, which can or not be ingested. With the powerful beak they break the seeds, even those with very hard stones.

White-eyed Parakeet

Reddish-bellied Parakeet

Who's this one ? I did not see it properly

It is also known that the parrots need to ingest clay during their lifetime, but what is this for? Studying the makaws in the Manu National Park (Peru), it was found that a great amount of seeds eaten by these birds contains toxic substances, such as alkaloids and tannins (see page 40). Once in the digestive system, the clay would absorb the toxins of the seeds to be later egested (they can also supply minerals).

Many parrots form couples for all their life and during the sexual courting they usually have beak contact, cleaning of feathers and in some species the male hands in food to the female.

Most of them build their nests in hollows of trees, as is the case of the species of Iguazú. They search for palm trees or other trunks which are decom-posing. As they use the hollow in the tree they save energy in the building of nests. They also ensure more stable temperatures than the exterior and a greater protection from storms and predators compared to open nests. A tree in a gap could be in a "neighbour-hood" well protected from predators, but it runs a greater risk of being torn down by a strong wind. It is also more convenient to have the hole away from the ground and the orientation of the entrance according to the prevailing winds is also very important. The competition for the spaces in the trees is great and the scarcity of these spaces can limit the building of nests.

The White-eyed Parakeet, the Reddish-bellied Parakeet, the Scaly-headed Parrot and the Blue-winged Parrot are species frequently observed in Iguazú.

From the paths of the Upper Circuit, at dusk one can see small flocks which are returning from their feeding areas to their roosting areas. In a great part of your walks you will have the chance of seeing these birds, unfailing inhabitants of the jungles.

Toucans

Toco-toucan

The Toucans belong to a family exclusive of America.

They move in flocks or small groups, as many fruit-eating birds. In the trees they usually move jumping and their legs, with two fingers facing backwards and two facing forward, provides them a good support.

They are dispersers of seeds (see page 34), but to know how well they do so it is important to take into consideration several factors. It will depend on the size of the fruit, how it is digested, or variables such as the time of the year. Thus in the breeding stage, part of the food carried to their young are fruits and their seeds will remain in the nest.

Their large bill is light, porous, with sharp or serrated borders, and they are generally of bright colours, in the outside and inside too. It allows them to reach fruit which are in branches which could not hold their weight, such as the Ambay.

The bill would also act as a warning signal and can be an effective threat to possible enemies, which can be large animals (as monkeys, birds of prey, snakes) or small ants. The naturalist Thomas Belt makes interesting comments in his observation of a toucan in Nicaragua which captured Army Ants (see page 72) from its nest with its bill to avoid them from invading its hollow.

The birds build nests in hollows which already exist in the trees and both parents take care and feed the offspring, which are born with a colourful beak which is shorter than the adult's.

In the Upper Circuit it is possible to find the Toco Toucan on trees in search of fruit or in small flocks, specially early in the morning or in the afternoon.

Their great beak allows an easy identification, even at a distance. You can also observe them in other sectors of your walks.

The Red-breasted Toucan, smaller than the former, but of a comparable beauty, is another of the five species of the family present in Iguazú.

tion. Their bill is finely serrated and their diet is based on fruit, although they also eat insects and several invertebrates.

They make their nests in the hollows of trees and in Iguazú there are records of fledglings of Surucua Trogons as from the month of July.

Tanagers and their Relatives

Fawn-breasted Tanager

Trogons

Surucua Trogon

The Surucua Trogon and the Black-throated Trogon inhabit the Iguazú jungle. They live on trees. Males have showy colours, although they pass unnoticed because of their static posi-

Names such as Blue-hooded Euphonia and Green-headed Tanager already tell us of the colouring and beauty of these restless birds. They generally move in groups and can be seen in search of food in plants as the Ambay, Guapoy, Caá-Votyrey (see page 38), or several epiphytes as the Cactaceae.

In your walks you might suddenly find flocks of several tanagers, euphonias and other species as the woodcreepers on a tree bearing fruits. They are

Magpie Tanager

Cotingas

The Cotingas are birds which feed almost exclusively on fruit, different to the already mentioned frugivores which supplement their diet with insects or other organisms.

They feed on the pulp of fruit and they generally regurgitate the seeds. The specialisation of some species is a good example of co-evolution between birds and fruit, that is why they are the best seed dispersers.

They have the advantage of having food available almost all year round, although their diet is not balanced, as it is poor in proteins. This could be the reason why the incubating period is

Red-ruffed Fruitcrow

noticeable, specially in winter, and can be formed by as much as 20 species.

The existence of these mixed flocks is explained as a strategy which can help them to find sources of food and to defend themselves from their predators, as "more pairs of eyes see better than one". It is known that some species give alarm signals.

With their short and robust bill, they obtain small fruits, buds, leaves, nectar of the flowers and some complement their diet with invertebrates.

There are species which make their nest on Bromeliaceae, and the Violaceous Euphonia, frequently found in Iguazú, has been seen using the abandoned nests of the Red-rumped Cacique (see page 94).

One of the larger species is the Magpie Tanager, easy to identify by its long tail and its black and white plumage.

very long, as in the case of the Bare-throated Bellbird (up to 30 days). This species does not nest in Iguazú, and there are few individuals registered in this Park.

The Cotingas include species of several sizes and in many cases the males are very colourful.
The Red-ruffed Fruitcrow is the largest representative, and it complements its diet with insects. It is a silent and lonely bird which usually perches in places which are not very visible. The fruit of some Lauraceae are important in their diet.

•••

Birds with a Varied Diet

Plush-crested Jay

Plush-crested Jay

The Plush-crested Jays belong to the family Corvidae. They move between trees in small, noisy groups and can imitate the sounds of several animals. They search for insects in the bark of trees, several invertebrates on the ground, they hunt while flying and they even eat some fruit and seeds, although they are not good dispersers. They usually introduce their closed bill in fruit or other foods to later open them with the help of the muscles of the jaw, as the caciques and cowbirds do, sometimes using their fingers to hold the food.

They build their nests on trees. Because of their abundance in Iguazú, their beautiful colours and trusting conduct, they are one of the birds which can be observed at a shorter distance and with a greater frequency.

Red-rumped Cacique and Giant Cowbird

The Red-rumped Cacique and the Giant Cowbird belong to the family Icteridae, a group which includes the orioles and blackbirds.

Most orioles and blackbirds have characteristic hanging nests, which ensure a greater isolation from predators. Although heavy rains and winds can loosen them from the branch they are hanging from, and, there are predators which can reach their eggs (as toucans and Brown Capuchin Monkeys in Iguazú).

The Red-rumped Cacique knits its nests with Pindó leaves fibres (see

Giant Cowbird

Red-rumped Cacique in its nest

makes it easy to identify.

It moves in noisy groups and feeds on fruit and several arthropods. The adaptation of the jaw muscles is amazing in this group of birds. It allows them to introduce the bill in a fruit and later open it to break the food. This unusual method is used also by the Plush-crested Jay.

Now imagine you are a mother and at any moment the following situation arises:

• Act one: you have a beautiful baby in a cradle.

• Act two: you have more than one beautiful baby in the cradle.

Something similar occurs to the Red-rumped Cacique when it finds eggs of the Giant Cowbird in its nest. This Cowbird can be seen in Spring and Summer in the gaps of the jungle, modified environments and in general on the ground.

Let's see what the relationship between both species is:

The female of the Giant Cowbird looks for a nest of the Cacique with eggs to invade it. In warm weather areas as in Iguazú, it is not strange to find the nests with eggs alone for a while; they do not need a continuous incubation as in cold climates.

Once in the nest, the Cowbird some-

page 54). In the Waterfalls area you can see a few samples of Pindó Palms with a great number of hanging nests, many of which are abandoned, as each year they build a new one. Generally the active nests are found close to the centre of the tree, where they would have more protection. The closeness to the presence of man would give them protection from other predators.

This Cacique is very common in Iguazú and its black plumage with a red rump times throws out the eggs of the other bird and later lays her own.

When the Red-rumped Cacique comes back, it most probably incubates the eggs and feeds the intruder fledglings, which develop faster than their own.

This is called brood parasitism and the Giant Cowbird does it with different species of caciques and orioles.

In the world there are nearly 100 brood parasite species of birds, including a species of duck. This is the Black-headed Duck, found in lagoons of Argentina's pampas.

On the other hand, there are caciques that build their nests close to the nests of some wasps and bees, while there are other birds who make their nests inside insect's nests, as ants or termites, to obtain protection.

This is the case of the Surucua Trogon which has been observed making its nest in aerial nests of termites; and of kingfishers, woodpeckers and parrots.

Scavengers Birds: Vultures

The Black Vulture is the most common carrion eater in the Waterfalls area. It has a roosting area in the trees which can be observed with certain proximity from the "Ventana" (window) of the San Martín Island and from the end of the Upper Circuit (Biguá Fall) with the help of binoculars.

The Turkey Vulture can also be found there. The showy King Vulture is another species found in Iguazú, although it is rare.

The vultures and their close relative, the Andean Condor, base their diet on dead animals. A unique biochemical mechanism allows them to be immune to botulin, a mortal toxin produced by the botulism bacteria in rotten meat.

Their featherless head is an adaptation to their type of feeding, as in large carcasses they introduce their head in the dead animal as they feed. When the head is exposed to the air, the sun helps in the killing of bacteria.

The Black Vulture can also ingest small

Black vultures

reptiles, eggs or bird fledglings, as could be from the small colony of the Black-crowned Night Heron, situated in the Lower Iguazú River.

They can be seen gliding in the Waterfall area, specially during the hottest hours, when the changes in temperature favour air currents formation. Their sharp eyesight allows them to find food from a great height. When there is a great number of vultures flying in a circle, they have probably found a dead animal.

The Turkey Vulture usually flies lower than its close relative, as it is guided by its sense of smell. In this way it can detect decaying bodies in the jungle, not visible from the air.

•••

Birds that Obtain their Food from Aquatic Environments

Black-crowned Night Heron and Snowy Egret

The Black-crowned Night Heron and the Snowy Egret are two species frequently found in the Waterfalls area.
The Black-crowned Night Heron has a small colony in the rocky coast of the Lower Iguazú River and can be seen from the footpaths of the Lower Circuit. This species makes its nest in Iguazú and both sexes build it, incubate and feed their young.

Why do they build their nests in colonies?
On one hand, there are several species of herons and other bird that do so. Generally it is a strategy used by seabirds.

The advantages of being in a group would be co-operation in the defence against attacks by predators, an efficient use of the limited areas where to build nests, to facilitate the search and to guide to places where food is available, or to increase the stimulus during the reproductive stage.
Particularly for the herons, all these variables may take place, except perhaps for the lack of space to build their nests.
Of course it also has its cons, because of the density of the colony there are confrontations due to continuous invasions of the territory, there can be

Black-crowned Night Heron

transmission of parasites or diseases or, for example, the stealing of materials to make nests.

The Snowy Egret also forms large colonies, but no nests have been found in the Park. They fly with their neck folded and their legs extended to the back.

With its long legs, the Snowy Egret wades the banks in search of food. It eats fish, frogs, tadpoles, aquatic invertebrates, and on the ground it feeds on small rodents and lizards.
Its long neck allows it to throw fast and precise attacks to trap its prey with its bill. The fish are ingested head first to avoid problems with their fins and scales.
The birds of prey are important predators of these birds, specially when young or as eggs in the colonies.

At dawn it is possible to see over the Waterfalls area the passing of elegant flocks of the Cattle Egret going back to their resting places.

Aninga

Olivaceous Cormorant and Aninga

These are two species which live and nest in Iguazú.
The Olivaceous Cormorant is more frequently found and can be seen alone or in small groups in the Inferior Iguazú River or in the Upper Iguazú River
It feeds mainly on fish. Under the water it swims with its neck slightly folded, to extend it in the precise moment and capture the prey with its beak. Once on the surface, it starts ingesting the fish, head first.
The Aninga is found in couples or alone and can be distinguished by its long and thin neck and its narrow bill and the plumage colours are different between sexes. It is more probable to observe it on rocks or branches close to the banks of the Upper Iguazú. It can be seen in the last part of the Upper Circuit or from the area of footbridges that lead to the Devil's Throat.

Fish, aquatic insect larvae, frogs, small turtles, lizards, snakes, baby caiman and some rats were found in the stomach content of Aningas.
Both the Olivaceous Cormorant and

the Aninga sometimes extend their wings when they come out of the water to dry them in the sun. They do not posses the uropygian gland with "oil" which aquatic birds place on their feathers using their bills in order to obtain impermeability. The extension of their wings also helps in the regulation of the body temperature, as there is more surface exposed to gain or lose heat, depending on what is needed.

Snail Kite

The Snail Kite is a gregarious bird of prey with a very specialised diet. Their habitats are aquatic environments, and

Snail Kite

in Argentina they are distributed from the North to Buenos Aires. In Iguazú they can be found during Spring and Summer. They can be seen looking for prey in the shallow waters (variable according to the river water level) from the footbridges leading to the Devil's Throat. Their hooked bill is amazing, adapted to a diet based on the *Pomacea* snails. The dark plumage of the males and its outstanding white rump and undertail coverts makes it unmistakable in its slow flight.

Dusk is a good moment to find them active. They capture their prey with their claws to later go to their "perches", generally a tree or bush where they take the snail out of its shell.

In Iguazú it is difficult to get close, but in other regions it is usual to observe the pile of empty shells they collect under their perches.

Both sexes take part in the nest building, making a new one each year. During the courting, sometimes the male offers a snail as a present to the female. She accepts it, takes the snail out of its shell to eat it and soon after the male mates with her. It has been proved that this mating mechanism related to feeding is an instinctive act and can be found in other species of birds, as some parrots.

As all birds of prey, the fledglings receive food from their parents and must remain some time in the nest till they are in conditions of flying. After this, they start in the search of food supply, a very difficult task for young inexperts, as the young Kingfishers know very well.

Kingfishers

Ringed Kingfisher

Kingfishers are quite lonely birds. They usually perch on branches of river banks or streams, ready to catch small fish, tadpoles or aquatic insects.

They dive submerging into the water in a fast way with their wings folded and emerge a short time after, many times with a prey in their robust bill.
As it usually happens with fish-eating birds (herons, egrets, cormorants), they ingest fish head first. Before swallowing the prey, they usually bang it against a branch. Sometimes, they remain in sta-

tic flight scrutinizing the water to suddenly dive as a dart in search of food.

During flight they emit a very characteristic sound, which resembles a rattle. They build their nests mainly in the river banks with the help of their powerful bill and with their legs they go throwing out the material. Both sexes incubate the eggs and feed the young. As days go by they accumulate "rubbish" in their nests and the parents must take a bath after their visit.

When the fledglings are about a month old they will start with the difficult task of hunting their own prey. They have been seen practising with leaves or small floating sticks, and they even take them out of the water and bang them against branches. In badly calculated trials in shallow areas they can crash into the bottom of the river.

It is possible to find these birds in environments such as the banks of the islands of the Upper Iguazú River during the walks on the footbridges going to the Devil's Throat, or downstream the Dos Hermanas Fall. The most frequent species is the Amazon Kingfisher, but there are also the Ringed Kingfisher and the Green Kingfisher. The difference in size would help to avoid competition among them in the search of food.

— • —

Mammals

In the Iguazú National Park 71 species of mammals have been recorded. The observation of most of them is improbable, although their footprints are signs of their presence.

Rodents and bats are the groups with a greater number of species. Opossums are also very well represented: only in the Waterfalls area 10 species have been identified (in Misiones there are 14 species known and 23 in Argentina).

The jungle's paths can be used by some mammals to move around. This is the case of the Agouti, brockets, foxes and cats as the Ocelot and the Jaguar. Species like the Tapir, the coatis or armadillos usually cross the paths to re-enter the dense vegetation.

In the walks through the Waterfalls Circuits, the Coati is the most frequently found mammal. The Macuco and Yacaratiá trails are good places to look for the Brown Capuchin Monkey, Agouti and Brazilian Squirrel.

Coati

The Coatis inhabit most habitats of the National Park. They are very active and since early in the morning they start moving in groups, scenting with their long snout in search of fruits, invertebrates and small vertebrates.

One of the individuals acts as a sentry and if there is any possible danger its alarm call would cause the dispersal of the other members of the group to the trees.

They can climb the trunks with the help of their nails, and if any of them is attacked on the tree, it would let itself fall to the ground. Their long and outstanding ringed tail helps them to keep balance.

In Iguazú they usually form groups of 5 to 10 individuals, although they can be more numerous. During Spring and Summer the females are pregnant, at the beginning of Autumn the offspring are born and in July the family groups start being seen.

Coati

The cats are their main predators.

In the Waterfalls area the coatis find plenty of food, specially in the garbage. They have changed their eating and behaviour habits respecting the natural environment of the jungle, where they form smaller groups.

Used to the human presence, their movement is restless and curious. This allows a great opportunity to observe them, but one must show indifference towards this mammal. It is your responsibility and good criterion not to feed them, in order not to modify their eating habits more and also to avoid possible risks. If food is offered to them, they remain in the place and they can even climb on people. With their sharp nails they can tear clothes or cause small injuries.

Brown Capuchin Monkey

The Brown Capuchin monkey belongs to the genus *Cebus*, a well distributed group in South America.

First of all, what does it mean to the monkeys to live on trees?
The main difference with the ground mammals is that they move in a three dimensional world. To move easily they must have a good eyesight, which measures distances precisely and its hand and feet must respond with equilibrium and co-ordination to catch on to the branches in the right moment after the jump.
On the other hand, their hearing is

very sharp to detect the members of their group and their enemies in a world covered by leaves where little can be seen.

Conclusion: They require a developed nervous system, which has gained complexity in the evolutionary lineage.

This is not all. The articulation of their arms includes shoulders with a unique mobility amongst the mammals and their hands have an opposable thumb which allows them to catch on to the branches. As if this were not enough, the Brown Capuchin and many other monkeys of the New World have a prehensile tail which hangs on to the branches and acts as a "safety belt". Although this species is not as specialised as other arboreal monkeys, where their tail acts as a fifth appendage and can be the only support.

Brown Capuchin Monkey

102

The Brown Capuchin generally moves in large groups and if you can observe them you will be surprised with their agility. In the area of the Macuco path a group of 24 individuals has been identified.

What do the Brown Capuchins eat?
Several fruit, seeds, insects, spiders, snails, eggs of birds and fledglings, frogs and other small vertebrates. It has been observed that when they eat fruit, they choose the ripest, and they seldom swallow them hole. They eat the pulp and throw away the seeds, acting as dispersers.

In Iguazú it was proved that this monkey obtains food from nearly 50 plant species, although the fruit of only a few species are dominant in their diet (see page 35). The Pindó palm offers a great source of food all year round, specially during the scarce Winter months. They obtain water from their food, but they can also take it from the caraguatáes or climb down the trees to get it from a spring.

Their jungle predators are large birds of prey and some cats, although in Iguazú- where there are few eagles- the pressure might be exerted only by cats.

The Macuco and the Yacaratiá trails are the best places to see the Brown Capuchin Monkeys. The noise of moving branches and their vocalisations will help you detect them. Their obser- vation can be more effective in Winter when many species of trees lose their leaves.

It is ideal to carry binoculars in these situations, to keep quiet and observe them. With their gestures and sounds they could be exchanging different messages, although we can not under- stand them.

After all it is considered that we, as human beings, when we are introduced to a person, only 10 % of our message is sent through words, 30 % is in the tones of our voice and 60 % in our body language, such as looks and ges- tures.

Agouti, Paca and Brazilian Squirrel

The Agouti and Paca are outstanding rodents of several jungle environments of America. Let's compare them. The Agouti is a diurnal mammal and, while you move in silence and are alert, you might be lucky and see it in the Macuco trail.
Instead, the Paca is larger and with its nocturnal habits, more difficult to find.

Both feed on fruits and seeds, but the Paca has a more varied diet as it feeds on leaves and also re-ingests part of its excrement to obtain more nutrients. The Agouti has a diet of fruits and seeds. As many of them are hard, it has been seen to sit down and with its

Brazilian Squirrel

Paca

Agutí

hands hold its "mouthful" to bite it, always ready to run away in case of danger. When there is abundant food, apart from eating it, it also buries seeds as storage. This behaviour might be more important in environments where there is a very dry season (with fruit scarcity) as in Costa Rica and Panama.

They have customary paths between feeding and sleeping areas and they leave olfactory signals.

They place seeds in several points, possibly to avoid the total loss in case they are found by predators as the peccaries. With this behaviour the Agouti acts as a good disperser (see page 33), as even

with their good memory, it hides more seeds than what it can recover.

The Brazilian Squirrel also bases its diet on fruits and seeds and it probably eats eggs. In Iguazú, the fruits of the Pindó palm are very important during Winter, when there is little food. It is very agile and active and has diurnal activity. It is found alone or in couples and can climb trees with great skill, using its long tail for balance and nails for anchorage.

• • •

The Bad Guys

We all know that the incredible "computerised" dinosaurs of films such as Jurassic Park form part of cinematographic achievements which fulfilled their aim: to impress in order to make us spend a good time and to make good profit. The same can be said of King Kong, or the mechanical monster of Jaws, among others.

Frugivorous bat

Pariparoba bush

Probably the vampires, as the sleepless Count Dracula, have appeared in more films. To such an extent that a lot of people associate bats only with "scary blood-sucking vampires" (except for the modern Batman).

The fact is that the special effects and the art of film making achieve incredible things. But not all is the "merit" of the creative people from this industry. Going back to ancient cultures, it is known that animals as sharks and snakes already originated in those days a mixture of fear and admiration.

In certain human groups and in some monkeys it has been proved that there is an innate aversion to snakes. But, from the respect and care one must have, to the terrible reputation which they have been given, there is a great difference.

Let's leave what the use of science fiction does to our perception of the animal life to penetrate the jungle once more.

There is a great diversity of bats, the only flying mammals of the world, mainly with nocturnal activity.

Some are efficient pollinators, others eat fruit and disperse seeds (see page 35), a few fly near the water to capture fish, some feed on birds and small frogs and toads, others feed on insects and of course there are some that feed on blood.

All these can be found in Iguazú, except for the pollinators and fruit-eaters, which are probably present but have not been captured yet.

The eye sight of these winged mammals is more or less developed according to the groups, and many have an efficient sonar system. They emit ultrasound signals whose echoing indicates a "terrain map" and the position of possible preys.

They have evolved from insectivore bats, and together with the Tyrants (see page 85) they are a good example of what is called "adaptive radiation": their "flexibility" has allowed them to cover many different niches.

Due to their night activity and their fast flight, their observation is difficult. Although you might not see them, you will know that the diversification of this fantastic group has made them play an important role in the jungle's dynamic.

— • —

The Weakness of the Strongest: Jaguar and Harpy Eagle

The Jaguar and the Harpy Eagle are the largest and most powerful hunters of jungle environments of America, one on the ground and the other in the air. Both travel wide territories on their own in search of their prey.

They hardly have any natural predator, they are fearful and efficient hunters and, at a first impression, any vertebrate which is smaller in size could be a good bite for them.
The Harpy can have a wing-span of two meters and the female weighs up to 9 kg; she is considered the most powerful bird of prey in the world. Its thick

legs with nails which can be 7 cm long, convert it in a predator capable of lifting preys of a considerable size.
It is known that they attack several animals on trees as monkeys, coatis, and Collared Anteater , but they also capture prey on the ground, as the Agoutis.

The Jaguar, with males that in Argentina can weigh 100 kg, is so agile and strong that it is the only hunter

Jaguar or Yaguareté

capable of confronting the Tapir. It swims well, can climb trees, and its camouflage colours (see page 47) allows it to surprise many prey.

The Jaguar diet includes peccaries, brockets, coatis and pacas and even caimans and smaller animals such as fish, birds, Tegu Lizard or turtles, and many others.

Its name in guaraní, yaguareté, means "real beast". Is a true symbol of the Misiones jungle. Everybody knows stories and legends about the great tiger, although very few have seen it.

Nothing can cause greater fear and respect in many creatures of the jungle as the roaring of the Jaguar or the shriek of the Harpy Eagle.

Never-the-less...

Being a large carnivore has its weak spots.
On one hand, they need wide territories and their niches are less abundant than those of a small animal, as a rodent. Thus, the Iguazú National Park is not large enough to protect a population of Jaguars. The green corridor project in Misiones allows large species to move in extensive territories (see page 111 and 125).

Although they hardly have competition with other species, they must compete among themselves to obtain a couple and in search of food. In small populations, the cost of each organism which remains "outside the system" is very high. For example, an infertile female, or in years of food scarcity (caused by environmental stress, such as droughts) they might not find enough food for their offspring.

Another important issue is that they need a long time to reach their sexual maturity and they have few offspring. The Jaguar takes three or four years before reaching sexual maturity. The females usually give birth to two cubs after a bit more than three months of pregnancy and, if her offspring survive, she will not reproduce the following year.

From research done in Guyana, it is known that the Harpy lays one or two eggs which is incubated for over two months. Only one fledgling survives. It is fed during five or six months before it is in conditions to leave the nest. It will continue receiving food during the first year of life and will reach sexual maturity when it is four or five years old. Only five nests of this species have been recorded in Misiones since 1989 (none in the Iguazú National Park, where there are no records of this eagle).

These species have become an easy target since the arrival of man due to their large size. The search for the hide of the Jaguar or the hunting of both species, as they are considered dangerous animals and enemies of domestic animals (plus the game hunting of the Jaguar), followed next.

The decrease of their natural areas is the main reason why they are endangered in Argentina.

The need of wide territories, the fact they have few offspring and their slow development are characteristics in common with most large mammals in the world (specially carnivores).

Thus tigers, rhinoceros, gorillas and other large animals are in danger of extinction.

Of course, their critical situation has been incited by man's search of some peculiarities, as the horns of rhinoceros, elephant's tusks or the hide of the Jaguar.

Surely, in most cases, the lack of their natural areas is their great crossroad. Now-a-days, they have all become "intensive care patients". Their destiny depends on the measures taken now and in the future.

The scientific knowledge for the managing and conservation of these species is scarce and hard to obtain. Whoever makes research on the Jaguar or the Harpy can spend many years of work to only see their studied animals in a few occasions.

With the Jaguars, the researcher must guide himself by footprints, excrement, remains of preys, and radio transmitters to follow their movements, or by interviews and surveys with local people. Very specialised animals, for example the Brazilian Merganser of Misiones, have also been driven to a critical situation by human activity.

The Jaguar and large eagles as the Harpy represent symbols of power and respect for the inhabitants of the Misiones jungle, ever since the Guaraníes dominated the territory. For many, they are transforming into symbols of the jungle we must conserve, including all the flora and fauna and the ecological processes which regulate it. Meanwhile, the surface area of the jungle keeps decreasing. And what is more serious, our society as a whole does not take real consciousness of this problem. Life in towns has many advantages, but it has no doubt disconnected us from the natural world.

To try to solve a problem, before anything else one must know exactly what the problem is.

Harpy Eagle

Bibliography

• Ackermann, J.D. 1996. Coping with the epiphytic existence: Pollination strategies. Selbyana Vol 9. Sarasota, Florida.

• APN-FAO. 1988. Plan de Manejo del Parque Nacional Iguazú. Proyecto, planificación y Gestión de los Parques Nacionales, Buenos Aires.

• Belt, T. (1911) 1928. The Naturalist in Nicaragua. J.M. Dent & Sons Ltd., New York.

• Benzing, D.H. 1996. The vegetative basis of vascular epiphytism. Selbyana Vol 9. Sarasota, Florida.

• Brewer, M.M. y de Argüello N.V. 1980. Guía Ilustrada de insectos comunes de la Argentina. Miscelánea, Nº 67. Inst. Lillo, Tucumán.

• Bucher, E. 1994. Posible impacto ambiental de obras para la atención de visitantes en el área Cataratas del P.N. Iguazú. Informe Técnico para la APN.

• Chebez, J.C. 1988 y 1990. La Selva Misionera I y II. Guías Educativas de Vida Silvestre. F.V.S.A.

• Chebez, J.C. 1994. Los que se van. Especies argentinas en peligro. Editorial Albatros.

• Chebez, J.C. 1996. Fauna Misionera. Editorial L.O.L.A.

• Collias, N.E. & Collias, E.C. 1984. Nest Building and Bird Behavior. Princeton University Press, Princeton.

• Conniff, R. 1996. Spineless wonders. Henry Holt and Company, New York.

• Crespo, J.C. 1982. Ecología de la comunidad de mamíferos del Parque Nacional Iguazú, Misiones. Rev. Mus. Arg. Cs. Nat., Ecología, Tomo III (2).

• De Vries, P.J. 1987. The butterflies of Costa Rica and their Natural History. Princeton University Press, Princeton.

• Dimitri, M.J.(Dirección); Volkart de Hualde, I.R.; Ambrosius de Brizuela, C.; Fano, F.A. (coautores). 1974. Flora arbórea del Parque Nacional Iguazú. Anales de Parques Nacionales Tomo XII, Buenos Aires.

• Emmons, L. 1990. Neotropical Rainforest Mammals. A Field Guide. The University of Chicago Press, Chicago.

• Erize, F.; Canevari, M.; Canevari P.; Costa, G.; Rumboll, M. (1981) 1995. Los Parques Nacionales de la Argentina y otras de sus áreas naturales. Incafo - Editorial El Ateneo.

• Fanelly, D. 1984. The Book of Bamboo. Sierra Club Books, San Francisco.

• Forshaw, J.M. 1973. Parrots of the World. Landsdowne Press, Melbourne.

• Fjeldsa J. & Krabbe N. 1990. Birds of the High Andes. Zoological Museum, University of Copenhagen.

• Forsyth A. & Miyata K. 1984. Tropical Nature. Macmillan Publishing Company, New York.

• Giai, A. 1976. Vida de un Naturalista en Misiones. Editorial Albatros.

• Heinonen Fortabat, S.; Schiaffino, K.; Bosso, A.; Oliva, A.; Marull, C.; Cervantes, R.; Mazar, J. y Acosta, S. 1994. Relevamiento Faunístico del área Cataratas Parque Nacional Iguazú. Descripciones de las comunidades, evaluación de la biodiversidad y detección de impactos actuales. Informe inédito. Delg. Téc. Nord. Argen. CIES-APN.

• Janzen, D.H. (Ed.) 1983. Costa Rican Natural History. The University of Chicago Press, Chicago.

• Johansson, D. 1974. Ecology of vascular epiphytes in West African rain forest. Acta Phytogeographica Suecica 59, Upsala.

• Johnson, A. 1996. Inventario Florístico del Parque Nacional Iguazú. Orquidaceae. F.V.S.A. (Programa Orquídeas). Informe Inédito F.V.S.A.- APN

• Kress, J. 1996. The systematic distribution of vascular epiphytes: an update. Selbyana Vol 9. Sarasota, Florida.

• Kricher, J.C. 1990. A Neotropical companion. Princeton University Press, Princeton.

• Laclau, P. 1994. La Conservación de los Recursos Naturales y el Hombre en la Selva Paranaense. Boletín Técnico de la Fundación Vida Silvestre Argentina.

• Leigh, R.G.; Rand S.A.; Windsor D.M. 1990. Ecología de un bosque Tropical. Smithsonian Tropical Research Institute, Balboa (Panamá).

• Malmierca, L.; Herrera J.; Schiaffino K.; Giorgis P.; Heinonen Fortabat, S. 1994. Relevamiento del área cataratas, Parque Nacional Iguazú, Informe de avance. Centro de Investigaciones Ecológicas Subtropicales (CIES), Del. Téc. NEA-APN. Cataratas del Iguazú.

• Marden, J.H. Enero 1990. Newton's Second Law of Butterflies. Natural History, New York.

• Margalef, R. 1986. Ecología. Ediciones Omega S.A., Barcelona.

• Meglitsch, P.A. 1978. Zoología de Invertebrados. Blume Ediciones, Barcelona.

• Moffet M.W. Julio 1995. Gardeners of the Ant World. National Geographic. Vol. 188, N°1, Washington, D.C.

• Munn, C.A. Enero 1994. Winged Rainbows Macaws. National Geographic. Vol. 185, N° 1. Washington, D.C.

• Murawski, D.A. Diciembre 1993. A Taste for Poison. National Geographic Vol. 184, N° 6, Washington, D.C.

• Narosky, T. & Yzurieta, D. 1987. Guía para la identificación de las Aves de la Argentina & Uruguay. Vazquez Mazzini Editores.

• Narosky, T. & Bosso A. 1995. Manual del Observador de Aves. Editorial Albatros.

• Norman, D.R. & Naylor, L. 1994. Anfibios y Reptiles del Chaco Paraguayo. Tomo I. Norman D., San José de Costa Rica.

• Pelt, J.M. 1985. Las Plantas. Biblioteca Científica Salvat, Barcelona.

• Placci, G.; Arditti, S.I.; Giorgis P.; Wutrich A. 1992. Estructura del "palmital" e importancia de *Euterpe edulis* como especie clave en el P.N. Iguazú. Yvyraretá 3.

• Preston, R. 1988. Butterflies of the World. MAFHAM, Facts of File Publications. New York.

• Putz, F.E. & Holbrook, N.M. 1996. Notes of the Natural History of Hemiepiphytes. Selbyana Vol 9. Sarasota, Florida.

• Rettig, N. Diciembre 1995. Remote World of the Harpy Eagle. National Geographic, Vol. 187, N° 2. Washington, D.C.

• Ridgely, R.S.; Tudor, G. 1994. The Birds of South America. Vol. II. University of Texas Press, Texas.

• Ruschi, Augusto. 1982. Beija-flores do Estado do Espírito Santo. Editora Ríos.

• Saibene, C.; Castelino, M. A.; Rey, N. R.; Herrera, J.; Calo, J. 1995. Inventario de las Aves del Parque Nacional Iguazú, Misiones, Argentina. L.O.L.A., Buenos Aires.

• Santos Biloni, J. 1990. Arboles autóctonos argentinos. Tipográfica Editora Argentina, Buenos Aires.

• Sick, H. 1985. Ornitología Brasileira, uma introducao. Vol. 1 y Vol. 2 Edit. Univ. de Brasilia, Brasilia.

• Short, L. 1982. Woodpeckers of the World. Delaware Museum of Nat. History, Greenville.

• Short, L. 1993. The Lives of Birds. American Museum of Natural History, New York.

• Schmit-Nielsen, K. 1984. Fisiología Animal, adaptación y medio ambiente. Ediciones Omega S.A., Barcelona.

• Simpson, G., 1982. The Book of Darwin. Washington Square Press, New York.

• Spichiger, R.; Ortega Torres, E.; Stutz de Ortega, L. 1989. Noventa especies forestales del Paraguay. Flora del Paraguay. Serie especial N° 3. Editions des Cons. et Jardin botaniques de la V. de Geneve & Missouri Bot. Garden (Saint Louis).

• Tudge, C. 1996. The Time before History. Scribner, New York.

• Wheelwright, N.T. & Janson C. 1985. Colors of fruit displays of bird-dispersed plants in two tropical forests. The American Naturalist, Vol. 126 N° 6.

• Wilson, E.O. 1996. In search of Nature. Shearwater Books, California.

The Misiones Jungle

After being amazed by the incredible relationships between the living organisms that inhabit the jungle, we have a clear idea of the complexity of this environment and how complicated it is to try to understand how it works. To cut down a tree in a natural environment does not only mean to obtain timber. It also means to leave homeless many species who use it as support, food and refuge, and to take with the enormous trunks part of the soil's nutrients which travel along the sap at the time of the cut-down.

The soil that maintains enormous trees of up to 30 or 40 m created the false idea of its fertility. The settlers believed: "If these trees can grow here, why won't my maize plants grow on this soil?" But the truth is that this soil, stripped from the green cover which protects it from the sun, wind and mainly the torrential rains, is washed away in a short period of time losing its thin fertile layer. That is why the poetic phrase that says that when: "one cuts a flower, the stars are moved" or the other phrase that says: "even the leaf that falls is counted by Nature" have in the jungle of Misiones their best example.

It is difficult to use the jungle without damaging it, consult it without frightening it, manage it without raping it. As a consequence, the people of Misiones have a dilemma. They posses a **green corridor** which joins the jungles of the Iguazú National Park and the Urugua-í Provincial Park, in the North, with the Yaborí Biosphere Reserve and the Moconá Provincial Park, in the South. Nearly 1,400,000 ha of almost continuous jungle, with the last chance of survival of superb components of our fauna, as the spotted Jaguar and the powerful Harpy Eagle. Most of the streams of Misiones which flow to the Alto Paraná River and the Alto Uruguay, which must be protected to ensure the supply of fresh water for the cities built on the banks of these rivers, appear in this area.

Luckily this immense area has a low population density, there is only one important city (San Pedro), thus it can be transformed into an International Biosphere Reserve. In this way, Misiones, Argentina and the world will have protected, in the long run, the survival of the last sample of this natural environment which once extended through the East of Paraguay, South of Brazil, and most of the province of Misiones, in Argentina.
Iguazú shall be then, not only the framework for the wonderful Waterfalls, but also the threshold of the mysterious kingdom of the jungle. It shall be the axis of the **green corridor** and the first natural reserve connecting three countries in South America, as it joins with Puerto Bertoni in Paraguay and the Iguaçú National Park in Brazil.
Only in this way shall we fully enjoy this environment; when we get to know that the "Ybirá Retá" (the land or country of trees) will be with us and marvel us for ever.

Juan Carlos Chebez
Director de la Delegación Técnica Regional NEA
Administración de Parques Nacionales
Av. Tres Fronteras 183 - C.C. 54 - (3370) - Puerto Iguazú - Misiones
(03757) 421-984/422-906 - E-mail: delegacion@parquesnea.com.ar

APN

Our National Parks

The National Parks of Argentina originated in 1903 when Francisco Moreno donated 7,500 hectares of his property in the region which would later be the Nahuel Huapi National Park. This donation was accepted by the National Government in 1904, forming the initial core of the Park, which was extended to 43,000 hectares in 1907, and completed in 1922 with a total surface of 785,000 hectares.

In the meantime, thanks to the negotiations done by Carlos Thays, the Argentine Government bought 75,000 ha in Iguazú, to form a National Park and a Military colony.

The consolidation of these two visionary men only happened in 1934 with the promulgation of the first National Park's Law, elaborated by Dr. Ezequiel Bustillo. Three years later, other areas were added to these parks: Lanín, Los Alerces, Perito Moreno and Los Glaciares.

This initial stage was characterised by a great impulse. Great investments in equipment and infrastructure aiming mainly at the development of tourism. This policy which was characterised by the protection of the areas of great beauty, changed with time.

Argentina is a country with an immense diversity of environments, many unique or shared in a small area with neighbouring countries. But a high percentage of these environments are not represented, or at least not enough, in the national protected areas.

The great agro-industrial development and the demographic growth of the end of the millennium, has lead to dramatic modifications in all natural environments, with the loss of genetic variability. This not only means its potential use, but also, in a great measure, the loss of the origin of the roots and culture of Argentina. That is why the Administración de Parques Nacionales has made a great effort in the concretion of new areas which represent and protect the immense native biological diversity of Argentina. The thirty three areas protected today, many recently created, preserve samples of many of these environments. Never-the-less, the total surface of these areas is hardly more than 1 % of the total area of Argentina. There are some biomes which are not represented, and several parks that are too small to ensure protection in the long run. This is the case of the Iguazú National Park, which shelters in its 67.000 hectares, more than 1.000 species of vascular plants, including 85 species of orchids among them; 430 species of birds, almost half of the bird diversity in Argentina. But a great amount of these biodiversity does not have viably populations within the park, and its therefore condemned if the deforestation and destruction of the environment continues. For this reason it is necessary to work together with the provinces and neighbouring countries, connecting natural areas and sharing experiences. The green corridor is a hope. There is still a lot to do.

Marcelo Canevari
Director de Interpretación y Extensión Ambiental
Administración de Parques Nacionales

Aves Argentinas and their work in Misiones.

Aves Argentinas has been developing since 1916 several tasks in favour of the knowledge and conservation of the birds of Argentina. Thanks to the annual contribution of nearly 1,500 members, our main goal consists in connecting the argentineans with their environment through the organisation of public information campaigns, divulgence magazines, such as Naturaleza & Conservación and Nuestras Aves, the organisation of congresses, an Argentine School of Naturalists, bird-watching courses and campaigns in all the country. We are also working on a program of Important Bird Areas, related to Birdlife International, world federation which we represent in Argentina. These activities are developed in Buenos Aires and a network of active members take the message of Aves Argentinas to all parts of the country.

Considering the fact that the Paranaense Forest is the most threatened biome in Argentina, we have placed special emphasis in conservation goals. We organized a group of enthusiastic collaborators, to be present in local newspapers, radios and television with conservationist messages; printed posters about dealing of fauna; gave courses for all ages; developed public information campaigns for relevant topics as the projects of dam building in the region. We also collaborate with the Provincial Ministry of Ecology raising funds for the purchase of fuel so the rangers can travel through the local protected areas. Parallel to this effort, we are working hard on the establishment of the first Centre of Recovery and Breeding of Threatened Birds, *Güira Oga*, which is a few kilometres from the Iguazú Waterfalls, inaugurated in 1997. In this reserve of 22 hectares with a good jungle coverage, belonging to the Ministry of Ecology and administered by Aves Argentinas, we have established a unique infrastructure to assist animals which have been confiscated or found hurt in the area. They are attended by professionals and after a meticulous work, the possibility of liberating them once more into the environment is evaluated. There are specially designed enclosures for birds, and a big quarantine room for the animals that have just arrived. The place is also an excellent centre of environmental education. Aves Argentinas sponsores another private reserve called *Yaguaroundi*, near San Pedro, in the heart of the Green Corridor.

Black-hawx Eagle
(photograph: Daniel Gómez)

Andrés Bosso
Executive Director, Aves Argentinas/AOP
25 de Mayo 749 2 6 (C1002ABO)
Buenos Aires, Argentina
54-11-4312-8958 / 1015 / 2284
info@avesargentinas.org.ar
www.avesargentinas.org.ar

AVES ARGENTINAS
Asociación Ornitológica del Plata

Centro de Investigaciones Ecológicas Subtropicales CIES

The Centro de Investigaciones Ecológicas Subtropicales, placed a few meters away from the Waterfalls, is the only centre of this type administered by the Administración de Parques Nacionales. It has a modest lodging for researchers who are working in the Iguazú National Park and its area of influence. Its objectives are:

• To incentivate and develop research and monitoring projects in the Iguazú National Park, with the purpose of defining general rules for the adequate protection of the ecosystem.

• To elaborate and propose research topics to develop in the Iguazú National Park. To co-ordinate the research program of the National Park.

• To contribute in the formation of human resources in ecology, biology and conservation, with special emphasis in the problems of the Subtropical Forests.

• To organise and maintain reference collections of flora and fauna of the Iguazú National Park; and a centre of bibliographic reference with specialised information on research, conservation and management of Subtropical Forests.

• To participate in the fulfilment of the environmental studies and reports planned for the Iguazú National Park.

CIES - Administración de Parques Nacionales
Av. Tres Fronteras 183 - (3370) - Puerto Iguazú - Misiones
(03757) 421-222

Let's conserve our roots:
The Fundación Vida Silvestre in the Rainforest of Misiones

The rainforest of Misiones is one of the most threatened ecosystems in Argentina, and at the same time, the one with greater biodiversity and complexity. That is why, from the very beginning, the Fundación Vida Silvestre Argentina has performed a consistent series of actions in favour of the effective protection and sustainable use of all its natural resources. The World Wildlife Foundation has included this environment in the 200 ecoregions in the world with priority for conservation. The FVSA and WWF have worked together promoting and designing a Trinational Strategy of Conservation of the Paranaense Forest with Argentina, Brazil and Paraguay, countries which share this biome. The problems affecting it and the capacity to implement possible alternatives that guarantee its perpetual conservation are analysed.

These are some of the programs, projects and actions developed by FVSA:
· Creation and implementation of the Urugua-í Wildlife Reserve, of 3,243 ha, together with the Alto Paraná company having a ranger in the area - with an off-road vehicle-doing research, control and publishing.
· A regional Office in Puerto Iguazú focused in research and conservation activities, one of the first documents achieved is the ecoregions vision.
· To negotiate the creation of the Urugua-í Provincial Park, of 84,000 ha, in compensation of the flooding produced by the Urugua-í Dam.
· Creation of the Wildlife Shelters by contacting owners of private rural establishments of high biological value.
· Study and conservation of the Misiones province orchids, specially in the Iguazú National Park.
· Preparation of didactic material and publications (posters, educational guides, pamphlets, audiovisuals, T-shirts, stickers, etc.) referred to the flora and fauna of the jungle.
· Divulging the value of conservation and the problems of the jungle of Misiones in mass media.
· Promotion of the Provincial Law which declares the Jaguar as a National Monument and help with species population studies.
· Supporting projects of other institutions: "Sustainable use of the Palmito Edible Palm", "Breeding of Pacas", "Preventive measures to avoid the attack of domestic cattle by cats" and "Effective instrumentation of the Protected Provincial Natural Areas".

**FUNDACION
VIDA SILVESTRE
ARGENTINA**

The Rainforest needs us all. Help us to protect it.
Oficina Central: Defensa 251 Piso 6º "K" (C1065AAC)
Buenos Aires, Argentina. Tel: (54 11) 4331-3631, Fax: extensión 24

Programa Selva Paranaense: Av. Córdoba 464 (N3370COQ)
Puerto Iguazú, Provincia de Misiones, Argentina.
Tel: (54 3757) 422370 vidasilvestre@arnet.com.ar

Don't lose argentinian falls!

Brasil
Argentina
Río Iguazú
Isla San Martín — San Martin Island

Garganta del Diablo — Estación Garganta del Diablo
Tren de la Selva — Jungle Train
Circuito Superior
Estación Cataratas
Viejo Hotel — Old Hotel
Circuito Inferior
Dos Hermanas
Sendero Yacaratiá
Sendero Macuco — Macuco Trail
Salto Arrechea

IGUAZÚ ARGENTINA

www.iguazuargentina.com

Full Moon	Lower Circuit	San Martin Island
Time 2:00 Hs — Distance 1.200 Mts	1:15 Hs — 1.700 Mts	2:00 Hs — 1.500 Mts
Visitor Center	Green Path	Upper Cicuit
25' — 150 Mts	20 — 600 Mts	1:15 Hs — 800 Mts

Don't lose argentinian falls!

Train of the Jungle

IGUAZÚ ARGENTINA ®

Carlos E. Enríquez S.A. y Obras U.T.E.
Concesionaria del Área Cataratas del Parque Nacional Iguazú

Casilla de Correo Argentino N° 24 - C.P. 3370 - Puerto Iguazú - Misiones - Argentina
TE(03757)491-469/470/476
info@iguazuargentina.com - www.iguazuargentina.com

Restaurant La Selva

"Seasonal Fruits" Chart

For practical reasons, periods are considered since the fruit is green till it ripens, even when it falls to the ground (there can be yearly variations).

Dimension: it refers to the length or greater diameter, in centimetres, of the structure (seed or fruit) which is indicated in green in the corresponding subtitle.

Fruit with small winged seeds

Common name	Dimension	Jan.	Feb.	Mar.	Apr.	May	Jun.	Jul.	Aug.	Sep.	Oct.	Nov.	Dec.
Azota caballo	0.7 cm	■	■	■									
Cedro misionero	0.5 cm							■	■				
Lapacho negro	0.5 cm								■	■			

Winged Fruit

Common name	Dimension	Jan.	Feb.	Mar.	Apr.	May	Jun.	Jul.	Aug.	Sep.	Oct.	Nov.	Dec.
Maria preta	6-7 cm								■	■			

Fruit like long pods

Common name	Dimension	Jan.	Feb.	Mar.	Apr.	May	Jun.	Jul.	Aug.	Sep.	Oct.	Nov.	Dec.
Anchico	7-15 cm							■	■				
Curupay	15-20 cm							■					
Ybirá-pitá	up to 10 cm									■			

Fruit with wings

Common name	Dimension	Jan.	Feb.	Mar.	Apr.	May	Jun.	Jul.	Aug.	Sep.	Oct.	Nov.	Dec.
Guatambú	3-4 cm							■					

Fruit with "propellers"

Common name	Dimension	Jan.	Feb.	Mar.	Apr.	May	Jun.	Jul.	Aug.	Sep.	Oct.	Nov.	Dec.
Guayaibí	3-4 cm	■											■

Fruit which, mainly after they fall, serve as animal's food

Common name	Dimension	Jan.	Feb.	Mar.	Apr.	May	Jun.	Jul.	Aug.	Sep.	Oct.	Nov.	Dec.
Timbó colorado	6-9 cm					■	■						

Fruit with seeds covered by a layer of pulp (aril)

Common name	Dimension	Jan.	Feb.	Mar.	Apr.	May	Jun.	Jul.	Aug.	Sep.	Oct.	Nov.	Dec.
Cancharana	2-3 cm							■	■				
Cupay	3 cm							■	■				

Fruit with fleshy pulp: preferred by several birds

Common name	Dimension	Jan.	Feb.	Mar.	Apr.	May	Jun.	Jul.	Aug.	Sep.	Oct.	Nov.	Dec.
Aguay	2-2,5 cm				■	■	■	■					
Ambay	2 cm	■	■	■	■	■	■				■	■	■
Cerella	1,5-2 cm	■	■										
Guapoy	1 cm	■	■										
Pindó	2-3 cm	■	■										
Caraguatáes	1 cm	■	■										
Güembe	1-2 cm	■	■										

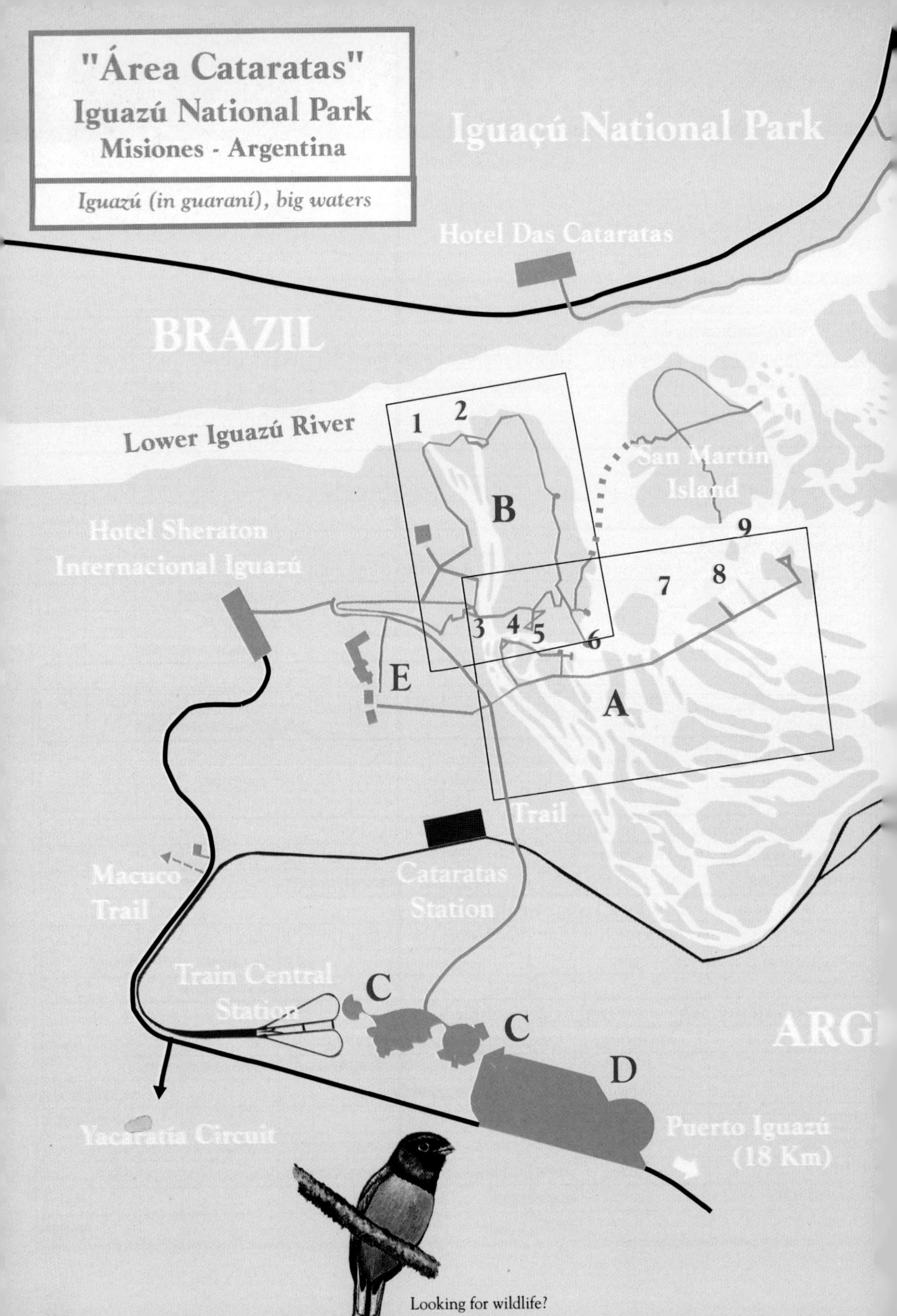

"Área Cataratas"
Iguazú National Park
Misiones - Argentina

Iguazú (in guaraní), big waters

Iguaçú National Park

BRAZIL

Hotel Das Cataratas

Lower Iguazú River

1 2

B

San Martín Island

9

Hotel Sheraton
Internacional Iguazú

7 8

3 4 5

6

E

A

Trail

Macuco
Trail

Cataratas
Station

Train Central
Station

C

C

D

ARGE

Yacaratía Circuit

Puerto Iguazú
(18 Km)

Looking for wildlife?
Try the trails.

Upper Iguazú River

Garganta del Diablo
Station

Iguazú National Park

PARQUE NACIONAL IGUAZÚ
ARGENTINA

References

━ **Vehicle roads**   ━ **Walking areas**   ━ **Railway tracs**

A Upper Circuit, see page 120.
B Lower Circuit, see page 122.
C Main Service Centre, see page 116: Tourist information, restaurant, fast food, ice-
   cream shop, souvenirs and handicrafts shops, telephone cabins, bathrooms,
   emergency assistance.
D Interpretation Centre: ecology of the rainforest, the native guaranies, the
   colonization and deforestation, and the project of the green corridor to preserve
   the forest.
E Jungle Explorer main office; restaurant and bar, see page 22.

   Each train station includes fast food services and bathrooms.

Some of the Main Falls (1-10, Argentina; 11-13, Brazil):

1	Lanusse	8	Adan y Eva
2	Alvar Nuñez Cabeza de Vaca	9	San Martín
3	Dos Hermanas	10	Unión
4	Chico	11	B. Constant
5	Ramirez	12	Floriano
6	Bossetti	13	Santa María
7	Mbiguá	U	UUUAAUHH !! Devil´s Throat

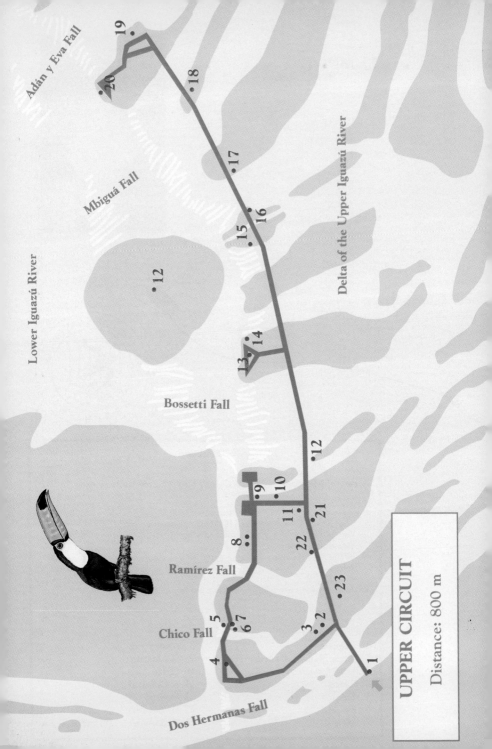

Lower Iguazú River

Adán y Eva Fall

Mbiguá Fall

Delta of the Upper Iguazú River

Bossetti Fall

Ramírez Fall

Chico Fall

Dos Hermanas Fall

UPPER CIRCUIT

Distance: 800 m

References of the Maps with Location of the Flora in the Waterfalls Circuits

The maps showing the flora of the Upper and Lower Waterfalls Circuits offer practical information. As an additional help in the identification, the respective references indicate the estimate heights of the plants (with approximate values every 5 m), a brief description of the place or characteristics of the tree and its outstanding epiphytes (see page 36). An indication of the page number remits to the corresponding text.

For better interpretation in both circuits, pay attention to:

• The shape of the leaves of the trees (see page 23) and the texture of their bark (see page 23); signs of attacks by herbivores (see page 40); the lianas (see page 38); several epiphytes, as the caraguatáes and Spanish moss (see page 57), Güembé (see page 59), Orchids (see page 60), Cacti (see page 37) and ferns (see page 56).

• With the help of the "Seasonal Fruits" Chart (see page 117) and the topic "Temptation of the Fruits", you can look for samples of the different strategies of dispersion of fruits and seeds in plants.

Key Word: Epiphyte (see page 36).

Upper Circuit

1) Azota caballo, 15 m. Wide crown. Its rough bark favours the growth of epiphytes: cacti, Orchids and small ferns. Intertwined lianas reach the top branches.

2) Cupay, 15 m. Dominant species in the Upper Circuit (see page 16). Specimen with oblique trunk. Epiphytes: outstanding group of orchids at an approximate height of 4m..

3) Mora blanca, 15 m. with two large branches which leave from a prostrated trunk.

4) María preta (see pages 33), 10 m. Grooved, straight trunk, long crown.

5) Ambay (see page 51), next to the bridge over the Chico Fall. Large, parasol leaves, thin and cylindrical trunk. The lichens are the dominant epiphytes on this species. Try to detect the Aztec Ants (for example, at the back of the leaves), and the nectaries.

6) Ubajay, 5 m. Riverine specimen, found behind a railing.

7) Cerella, 5 m (see page 34). its bark peels off (characteristic of Mirtaceae). Epiphytes: Ribbon-like cacti.

8) Cupay, 20 m. Two specimens with a straight trunk, branched in the upper part, growing between Pindó Palms.

9) Inga, 5 m.

10) Cupay, 20 m. Notice the reddish colour underneath its bark (characteristic of this species). To its right, an Ambay can be seen.

11) Mora blanca, 15 m. Branched from the base, wide crown.

12) Tacuaruzú Canes (see page 55), on islands.

13) Cupay, 20 m.

14) Curupay, 20 m. It is a good opportunity to compare the shape of the crown, leaves and fruit with the Cupay of 13).

15) Curupay, 15 m. Riverine. Straight trunk, branched in the upper part. Outstanding epiphytes: Cacti and Güembé with roots that descend surrounding the trunk; large caraguatáes.

16) Mborebí caá-guazú, 5 m. Specimen found just in front (and on the other side of the footbridge) of the homage monolith of the ranger Bernabé Mendez. Outstanding epiphytes: ribbon-leafed ferns (see page 56), orchids, cacti; hemiparasites: Caá-votyrey (see page 38). Behind, a riverine specimen of Azota caballo.

17) Ombú-ra, 5 m. Riverine. Wide crown, very branched. Large leaves, good observation from the footbridge. Epiphytes: small claveles del aire, orchids, small caraguatáes (*Vriesea* sp.).

18) Curupay, 5 m. Good observation of leaves and pods. Dense cover of epiphytes: Orchids, Cacti; caraguatáes (*Aechmea* sp., with greyish, serrated leaves ; *Vriesea* sp., smaller, with yellowish-green leaves with smooth edges.

19) Curupay, 10 m. With trunk covered with thorns, a frequent characteristic of young trees of this species.

20) Anchico, 15 m. Trunk and main branches almost completely covered with epiphytes: orchids, Cacti, caraguatáes, ferns, mosses.

21) Cancharana, 15 m (see page 34).

22) Carayá-bola, 10 m. Riverine specimen. Good observation of its fruits and large compound leaves, attacked by herbivores quite a lot (see page 40).

23) Mora blanca, 15 m; with Guapoy (see pages 52) growing on it. A few meters from the main path, in a gap of the jungle. Epiphytes: Güembé and a large Caraguatá.

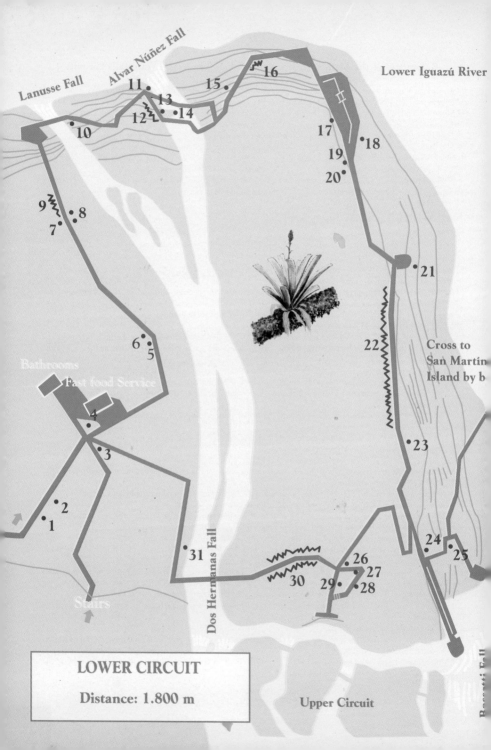

Lanusse Fall

Alvar Núñez Fall

Lower Iguazú River

11

15

16

13

12 14

10

17

18

19

20

9

8

7

21

22

Cross to
San Martín
Island by b

6

5

23

Bathrooms

Fast food Service

4

3

24

2

25

1

31

26

27

30

29

28

Dos Hermanas Fall

Stairs

LOWER CIRCUIT

Distance: 1.800 m

Upper Circuit

Lower Circuit

1) Young Guapoy, "strangling" host tree.

2) Aguay, 15 m. Epiphytes: large caraguatá and Cacti. Fruit growing directly on the branches (see page 31).

3) Ubajay, 15 m (see page 32). Its bark peels off, a characteristic feature of the Mirtaceae; orange trunk.

4) Aguay, 20 m.

5) Alecrín, 20 m (see page 43).

6) Yacaratiá, 15 m. Next to the Alecrín, parasol leaves, resemble those of the Ambay (but smaller).

7) Guayaibí (see page 33), 20 m. Close to the path. Placed almost in front of the Pindó Palms in 8). Observe the adventitious vines (see page 38) stuck to its bark.

8) Pindó Palms (see page 54). Observe the scarcity of flowering epiphytes on palms, although the lichens dominate on the trunks (with several colours, as the different shades of pink).

9) Mborebí caá, 3 m, bush with beautiful bluish flowers (they blossom in September).

10) Azota caballo, 10 m.

11) Lying Cacti, with thorns (see page 37). The epiphyte Cacti of Iguazú do not have spines, probably because they are less predated on trees.

12) Area with large rocks (under conditions of shade and high humidity). Watch the *Peperomias* (heart-shaped leaves), small ferns, orchids, Cacti, caraguatáes, etc. growing of on them. This is showing that when they grow on trees they are not parasites, as they can grow perfectly well on rocks.

13) Lapacho Negro, 20 m. Isolated specimen, found just before the upper balcony of the Alvar Nuñez Cabeza de Vaca Fall. Long and straight trunk.

14) Good observation of the Güembe (see page 59).

15) Ybirá-pitá, 15 m (see pages 27, 32). Straight trunk, with a Pindó Palm very close to it.

16) Thicket of chaguares or Ihvirá (see page 59). Compared to the caraguatáes, their leaves are more coriaceous (leather-like texture) and serrated and they do not form the evident water tank. Beautiful pink flower-head.

17) cancharana, 10 m.

18) Mora blanca, 20 m. Wide crown,

19) Carayá-bola, 10 m.

20) Cancharana, 20 m. Long and wide crown.

21) Anchico, 20 m. Wide crown, many epiphytes.

22) Shady and very damp area. Many *Peperomias* (heart-shaped flowers), small ferns, debris (see page 16).

23) Peteribí (see page 33), 15 m. Some 4 m in front there is a large tree next to the footbridge.

24) Amazing lianas "strangling" an Anchico's trunk. It occurs due to the growth of the tree in width.

25) Carayá-bolá, 20 m. Wide crown. Large specimen found between the rocks, halfway from the descent to the lower balcony of the Bossetti Fall.

26) Peteribí, 20 m.

27) Rabo, 15 m.

28) Alercrín, 25 m.

29) Canelón, 20 m. Long and straight trunk. Amazing "lenticels" of pink cork on it.

30) Area with Pariparoba bushes (see page 35).

31) Anchico, 20 m. The most noticeable epiphytes are large caraguatáes.

E

Iguazú area references

A. Área Cataratas (see page 118)
B. Puerto Iguazú (see page 126)
C. Foz de Iguaçú
D. Ciudad del Este
E. Itaipú Dam

Iguazú National Park
1. Hotel Sheraton Internacional Iguazú

Around Puerto Iguazú
1. Iguazú Grand Hotel - Resort & Casino
2. Hotel Cataratas
3. Hotel Las Orquídeas
4. Hotel Tropical
5. Hotel Margay
6. Bungalows and Camping Viejo Americano

1. Güira Ogá
2. Aripuca
3. Horse-back riding

D

Paraná River

C

Brazil

International Airport

Iguaçú National Park
170,000 ha
Subtropical Paranaense Rainforest

B 1 3
  5
2 4 6
2 3
1

Paraguay

Iguazú National Park
67,000 ha
Subtropical Paranaense Rainforest

Iguazú River

Argentina

RN 12

1

A

Enjoy Nature,
every place,
every moment.

Posadas
297 Km

International
Airport

RN 101

Paranaense Subtropical Rainforest
Misiones - Argentina (see page 111)

P. N. do Iguaçu

P. N. Iguazú

BRAZIL

GREEN CORRIDOR

PARAGUAY

Moconá

MISIONES
ARGENTINA

LANDSAT V satellite images from Misiones Province and
neighbouring countries (Paraguay and Brazil).
October, 1996.
Green tones: Native Jungle and forestations.
Ministerio de Ecología y R.N.R., Pcia. de Misiones.

Puerto Iguazú

Tancredo Neves International Bridge

Brazil

Iguazú River

Paraguay

Paraná River

Argentina

Parakeets around, the Jungle is close.

Custom Area

Iguazú Grand Hotel Resort & Casino

Raíces Argentinas

To the Falls and the Jungle

San Lorenzo 19

El Urú 12

Guaraní 7

Belgrano

Beltrán

J.P. Amarante

Córdoba Av.

Misiones Av.

Brasil Av.

Moreno

Bompland

Victoria Aguirre Av.

República Argentina

Los Cedros

Hito Argentino 7

Paseo de Artesanos

Tres Fronteras Av. 1

Nahú

Corrientes

APN

PUERTO IGUAZU REFERENCES

1. Hotel Esturión
2. Hotel Latino
3. Hotel Saint George
4. Hostería Los Helechos
5. Hotel Alexander
6. Hotel El Libertador
7. Hostería Casa Blanca
8. Pirayú Tiempo Compartido
9. Residencial La Cabaña (Hostelling International)
10. Hotel Misiones
11. Hotel Tierra Colorada
12. Residencial Lilian
13. Residencial Paquita
14. Residencial San Fernando
15. Residencial Uno (Hostelling International)
16. Residencial Uno (Hostelling International)
17. Hotel Bompland
18. Residencial Arco Iris
19. Residencial Río Selva
20. Residencial San Diego
21. Hostel Iguazú Falls

Where to eat

1. La Rueda
2. El Charo
3. El Tío Querido
4. Jardín Iguazú
5. Fechorías
6. La Barranca, bar
7. La Reserva, bar

Museums

Imágenes de la Selva

Mbororé

raíces argentinas

PASEO DE COMPRAS

TAKE YOUR CHOICE, WE OFFER THE OPTIONS
FOR YOUR NEEDS:

*Handicrafts, woodcrafts, leathergoods, silverwear,
potteries, jewelry.*

*Also T-shirts, postcards, books, yerba mate, wines, cheeses
and regional jams.*

Telephones and post office available.

Ruta Nacional 12 - Km 1640 - Puerto Iguazú - Misiones, Argentina
Ph. 54- 3757- 423635

FCB

IF THE FALLS

WEREN'T

SO CLOSE,

it could

be a cruiser.

When you arrive at the Iguazú Grand Hotel, you will enjoy several sensations at the same time. The elegance of each one of the places, the pleasure of the unique casino of Iguazú, the comfort of all our suites, the quiteness that grants our security to you, the excellent service recognized at a world-wide level in our restaurants and bars, in addition to the tennis courts, the swimming pools and our recreation department. For all these reasons, if you are looking for elegance, pleasure and entertainment, you know which is the only place in Iguazú where you can find them together.

Iguazú Grand Hotel
Resort & Casino

Buenos Aires (54-11) 4372-8700 – info@iguazugrandhotel.net – Reservations: 0800-4-IGUAZU – reservas@iguazugrandhotel.net – www.iguazugrandhote